ADVANCES IN

GEOPHYSICS

VOLUME 41

Advances in
GEOPHYSICS

Edited by

RENATA DMOWSKA

Division of Applied Sciences
Harvard University
Cambridge, Massachusetts

BARRY SALTZMAN

Department of Geology and Geophysics
Yale University
New Haven, Connecticut

VOLUME 41

ACADEMIC PRESS
San Diego   London   Boston   New York
Sydney   Tokyo   Toronto

This book is printed on acid-free paper.

Copyright © 1999 by ACADEMIC PRESS

All Rights Reserved.
No part of this publication may be reproduced or transmitted in any form or by any means, electronic or mechanical, including photocopy, recording, or any information storage and retrieval system, without permission in writing from the Publisher.
The appearance of the code at the bottom of the first page of a chapter in this book indicates the Publisher's consent that copies of the chapter may be made for personal or internal use of specific clients. This consent is given on the condition, however, that the copier pay the stated per copy fee through the Copyright Clearance Center, Inc. (222 Rosewood Drive, Danvers, Massachusetts 01923), for copying beyond that permitted by Sections 107 or 108 of the U.S. Copyright Law. This consent does not extend to other kinds of copying, such as copying for general distribution, for advertising or promotional purposes, for creating new collective works, or for resale. Copy fees for pre-1999 chapters are as shown on the title pages. If no fee code appears on the title page, the copy fee is the same as for current chapters.
0065-2687/99 $30.00

The article by Steven C. Cohen titled "Numeric Models of Crustal Deforrmation in Seismic Zones" is a U.S. Government Work in the public domain.

Academic Press
a division of Harcourt Brace & Company
525 B Street, Suite 1900, San Diego, California 92101-4495, USA
http://www.apnet.com

Academic Press
24-28 Oval Road, London NW1 7DX, UK
http://www.hbuk.co.uk/ap/

International Standard Book Number: 0-12-018841-4

PRINTED IN THE UNITED STATES OF AMERICA
99 00 01 02 03 04 BB 9 8 7 6 5 4 3 2 1

CONTENTS

CONTRIBUTORS vii

Oscillatory Spatiotemporal Signal Detection in Climate Studies: A Multiple-Taper Spectral Domain Approach

MICHAEL E. MANN AND JEFFREY PARK

1. Introduction	1
1.1 Motivation and Overview	1
1.2 Signal and Noise in Climate Data: Dynamical Mechanisms	3
2. Traditional Methods of Oscillatory Climate Signal Detection	6
2.1 Signal and Noise Assumptions: A Synthetic Data Set	7
2.2 Conventional Approaches to Signal Detection	17
3. MTM-SVD Multivariate Frequency-Domain Climate Signal Detection and Reconstruction	30
3.1 Signal Detection	31
3.2 Signal Reconstruction	34
3.3 Testing the Null Hypothesis: Significance Estimation	36
3.4 Application to Synthetic Data Set	41
3.5 Effects of Sampling Inhomogeneities	45
4. Applications of MTM-SVD Approach to Observational and Model Climate Data	49
4.1 Global Temperature Data	51
4.2 Northern Hemisphere Joint Surface Temperature and Sea-Level Pressure Data	72
4.3 Long-Term Multiproxy Temperature Data	101
4.4 Seasonal Cycle: Observations vs CO_2-Forced Model Simulations . .	111
5. Conclusion	121
Acknowledgments	122
References	123

Numerical Models of Crustal Deformation in Seismic Zones

STEVEN C. COHEN

1. Introduction	133
1.1 Elastic Half-Space Model of the Earthquake Cycle for an Infinitely Long Strike-Slip Fault: An Illustrative Model	137

2. Coseismic Deformation 139
   2.1 Elastic Dislocation Theory 139
   2.2 Early Applications of Dislocation Theory 145
   2.3 Extensions beyond the Uniform Elastic Half-Space Model 152
   2.4 The Finite-Element Method 158
3. Time-Dependent Effects 162
   3.1 Deep Fault Creep 162
   3.2 Viscoelastic Flow 163
   3.3 Kinematic Models of the Entire Earthquake Cycle 178
   3.4 Other Issues 189
4. Crustal Deformation near Specific Faults 204
   4.1 Nankai Subduction Zone, Japan 204
   4.2 Eastern Aleutian Subduction Zone, Southcentral Alaska 210
   4.3 San Andreas Transform Fault System, California 214
5. Epilogue 217
   Acknowledgments 218
   Appendix 218
   References 222

   INDEX 233

CONTRIBUTORS

Numbers in parentheses indicate the pages on which the authors' contributions begin.

STEVEN C. COHEN (133), Geodynamics Branch, Goddard Space Flight Center, Greenbelt, Maryland 20771.

MICHAEL E. MANN (1), Department of Geosciences, University of Massachusetts, Amherst, Massachusetts 01003.

JEFFREY PARK (1), Department of Geology and Geophysics, Yale University, New Haven, Connecticut 06520.

OSCILLATORY SPATIOTEMPORAL SIGNAL DETECTION IN CLIMATE STUDIES: A MULTIPLE-TAPER SPECTRAL DOMAIN APPROACH

MICHAEL E. MANN

Department of Geosciences
University of Massachusetts
Amherst, Massachusetts 01003-8520

JEFFREY PARK

Department of Geology and Geophysics
Yale University
New Haven, Connecticut 06520-8109

1. INTRODUCTION

1.1. Motivation and Overview

In order to properly assess the potential impact of forcings external to the climate system (e.g., possible anthropogenic enhanced greenhouse forcing), it is essential that we understand the background of natural climate variability on which external influences may be superimposed. Atmosphere–ocean–cryosphere interactions include many feedbacks that have time scales of years and longer. These feedbacks can, in principle, lead to irregular, but roughly cyclic, low-frequency climate variations (perhaps the most well-known example of which is the El Niño/Southern Oscillation or "ENSO"). If we can separate, in historical and proxy climate data, large-scale oscillatory, interannual and longer-period climate "signals" from the "background" climate variability, (1) it becomes easier to distinguish natural climate fluctuations from presumed anthropogenic or other external (e.g., solar) effects; (2) dynamical mechanisms potentially inferred from these signals provide a means of validating numerical climate models; and (3) these signals can themselves potentially be used for long-range climatic forecasting.

The complex behavior of the climate system challenges any single exploratory data analysis method. Nonlinear dynamical mechanisms, for example, could connect variations on widely differing time scales. Some truly episodic phenomena, such as climatic responses to large volcanic eruptions, seem best suited for study in the time domain. Others, such as the periodic changes associated with the seasonal temperature in surface temperatures, are better suited for study in the frequency domain. For certain phenomena, it is not clear whether an oscillatory or an episodic picture is most appropriate. For example, both the statistical model of a step-wise discontinuity in the climate during the latter 1970s (e.g., Trenberth, 1990) and that of oscillatory behavior with a particularly abrupt variation in climate regimes occurring at about that time (Latif and Barnett, 1994; Mann and Park, 1994; 1996a, b) but with similar analogues at other times (e.g., 1900, 1915, 1940, 1955—see Mann and Park, 1994; 1996a, b) have been used to describe large interdecadal fluctuations in the North Pacific and Northern Hemisphere climate in recent decades. Depending on the null hypothesis and statistical criterion employed, both statistical models can be argued for at reasonably high levels of confidence. Still other phenomena, such as ENSO, exhibit a mix of time-domain, or "event," characteristics and frequency-domain, or "oscillatory," characteristics. Later in this introduction, we present a skeletal overview of the potential dynamical mechanisms behind low-frequency climate variability. In later sections, we present attempts to isolate and reconstruct "quasi-oscillatory" components of the climate system, with characteristic interannual-to-century time scales, using a powerful multivariate statistical technique called MTM-SVD. This technique combines the multiple-taper spectrum estimation methods (MTM), developed by Thomson (1982), with a principal-components analysis using the singular-value decomposition (SVD). Section 2 presents more traditional methods for oscillatory climate signal detection.

MTM-SVD detects an oscillatory signal in a spatially distributed data set (e.g., gridded historical climate fields) by identifying an unusual concentration of narrowband data variance in a particular large-scale pattern, relative to the random fluctuations of the background climate variability. Although formulated and applied as a signal detector in the frequency domain, MTM-SVD can be used to reconstruct the time history of any potential oscillatory climate signal, as well as its spatial pattern. Secular trends in the data can also be detected and reconstructed in the MTM-SVD approach through a treatment of the near-zero frequency data variance. The evolution of an oscillatory signal over time, either in amplitude, frequency, or spatial pattern, can also suggest secular changes in the climate system, caused either by long-term natural variability or by possible external forcings.

A proper estimate of the statistical significance of putative oscillatory signals is crucial in the application of MTM-SVD. As with most multivariate techniques, statistical inference is most straightforward if the null hypothesis for background variability is simply specifiable (e.g., spatially uncorrelated white noise). Climate data falls far short of this ideal, as its random fluctuations exhibit significant correlations in both space and time. Much of Section 3 describes numerical experiments that demonstrate how to adjust the confidence levels for signal detection in MTM-SVD to account for such correlations. Section 3 describes the MTM-SVD method both formally, building upon the conceptual framework of other time-series methods, and by demonstration on a variety of synthetic data sets.

The MTM-SVD methodology has been used for signal detection and reconstruction in global temperature data over the past century (Mann and Park, 1994), joint fields of surface temperature and sea-level pressure in the Northern Hemisphere (Mann and Park, 1996b) and their relationship to continental hydroclimatic variations in North America (Mann et al., 1995a), low-frequency signals in the Atlantic and Pacific oceans (Tourre et al., 1997), signal detection in global (Mann et al., 1995b) and regional (Bradley et al., 1994; Rajagopalan et al., 1996) long-term climate proxy networks, the analysis of radionuclide tracers of the atmospheric general circulation (Koch and Mann, 1996), long-range climatic forecasting (Rajagopalan et al., 1998), and model vs observational "fingerprint detection" of anthropogenic forcing (Mann and Park, 1996a). We demonstrate in Section 4 how MTM-SVD can be applied to some of these issues.

1.2. Signal and Noise in Climate Data: Dynamical Mechanisms

In a crude approximation, background climate variability can be described by a simple "red-noise" model, in which the thermal inertia of the slow-response components of the climate system (e.g., the oceans as well as the cryosphere) tends to integrate the approximately white-noise forcing provided by weather systems. This process leads to enhanced noiselike variations at progressively longer periods (Hasselmann, 1976). More detailed noise models have been developed which take into account the additional effects of convective and diffusive exchanges between the mixed layer and the deeper ocean (Wigley and Raper, 1990). More generally, climatic noise can be characterized as exhibiting a "colored-noise" spectrum, associated with some underlying spatial correlation structure. Such "noise," however, is insufficient to describe the natural variability of the climate that arises from internal oscillatory modes of the climate system which are either self-sustained through nonlinear dynamics or stochastically excited by the noise itself. These low-frequency modes or "signals" may further compound the detection of anthropogenic climate forcing (see, e.g., IPCC, 1996; Barnett et al., 1996). The identification of such signals may have a profound importance in its own right, providing the possibility of skillful climate forecasting at decadal and longer lead times (see Latif and Barnett, 1994; Griffies and Bryan, 1997; Rajagopalan et al., 1998). For both reasons, the detection and description of low-frequency oscillatory climatic signals represents a problem of paramount importance both scientifically and societally.

Besides the seasonal and diurnal cycles, there is scant evidence for truly periodic climate signals (see Burroughs, 1992). Many climatic processes, nonetheless, appear to exhibit some oscillatory character, describing spatially coherent climatic variations which tend to oscillate between different states owing to a variety of possible linear or nonlinear feedback mechanisms. Such "quasi-oscillatory" signals, as we term them, are marked by a dominant time scale of variation, and often by finite, somewhat episodic, spells of large-amplitude oscillation. Perhaps the best-known example is the El Niño/Southern Oscillation which exhibits oscillatory variability within a 3 to 7-year time scale range, apparently further organized into distinct low-frequency (4- to 6-year) and high-frequency (2 to 3-year) narrow frequency bands (e.g., Barnett, 1991; Keppenne and Ghil, 1993; Dickey et al., 1992; Ropelewski et al., 1992; Mann and Park, 1994; 1996b). Such behavior can be associated with underlying coupled ocean–atmosphere dynamics which are presently understood at a reasonably fundamental level (Cane et al., 1986; Philander, 1990). There is mounting evidence both from observational analyses and from a variety of theoretical climate model investigations that similar types of oscillatory signals may exist in the climate on decadal-to-century time scales. Several workers have isolated decadal-to-interdecadal [e.g., Folland et al., 1984; Ghil and Vautard, 1991; Allen et al., 1992; Mann and Park, 1993; 1994; Royer, 1993; Mann et al., 1995a, b; Dettinger et al., 1995; Mann and Park, 1996a, b) and more speculative century-scale (Schlesinger and Ramankutty, 1994; Mann and Park, 1994; Mann and Park, 1996a, b) oscillatory behavior in instrumental climate records spanning a little more than the last century. The investigation of longer-term proxy data supports the existence of interdecadal (Mann et al., 1995b) and century-scale (Stocker et al., 1992; Mann et al., 1995b) climate signals prior to the twentieth century.

While many studies have attributed observed decadal-to-century-scale variability to external forcing due to the 18.6-year soli-lunar tide (e.g., Mitra et al., 1991; Currie and O'Brien, 1992; Royer, 1993), the ~ 11-year solar cycle (e.g., Labitzke and van Loon, 1988; Tinsley, 1988; Mitra et al., 1991; Currie and O'Brien, 1992) and its 22-year subharmonic or "Hale" cycle (Vines, 1986), and low-frequency changes in solar irradiance forcing (e.g., Friis-Christensen and Lassen, 1991, Lean et al., 1995), the most plausible oscillatory mechanisms—both in terms of physical mechanisms and in terms of their similarity in character to observed patterns of variability—involve natural oscillatory processes of the ocean or the coupled ocean–atmosphere system. A convenient categorization of possible mechanisms is provided by Stocker (1996), including (a) the interaction between the meridional overturning "thermohaline" circulation and the wind-driven circulation (Weaver and Sarachik 1991; Weaver et al., 1991; Huang, 1993; Cai and Godfrey, 1995); (b) the interaction of thermally generated baroclinic gyre anomalies and the thermohaline circulation (Delworth et al., 1993; Greatbatch and Zhang, 1995); and (c) the basin-scale advection of surface salinity (Maier-Reimer and Mikolajewicz, 1989; Mysak et al., 1993; Griffies and Tziperman, 1995; Schmidt and Mysak, 1996) or temperature (Saravanan and McWilliams, 1995) anomalies influencing deep water production and meridional overturning. A fourth category, not highlighted by Stocker (1996) but which has nonetheless gained recent prominence, involves the gyre-scale advection of thermal anomalies in the Pacific basin, associated changes in the thermal structure of the upper ocean, and its feedback on the atmospheric windstress profile (Latif and Barnett, 1994; Von Storch, 1994). In addition, other studies have suggested that the interaction between high-latitude brine release and ocean circulation (Yang and Neelin, 1993), subharmonic generation arising from the nonlinear interaction of sea-ice and high-latitude heat/freshwater fluxes (Yang and Huang, 1996; see also Saltzman and Moritz, 1980 and Saltzman, 1982 for a more general discussion of the underlying nonlinear dynamics) when driven by an annual cycle, and the interaction between ice-cover and thermal insolation (Zhang et al., 1995), coupled arctic sea-ice/atmospheric circulation processes (Mysak and Power 1992; Darby and Mysak, 1993), may lead to organized variability on decadal-to-century time scales. A final possibility is that such variability is simply the product of the fundamentally chaotic interaction of the atmosphere and the ocean–atmosphere system at decadal and longer time scales (see Lorenz, 1990; Roebber, 1995; Liu and Opsteegh, 1995; Kurgansky et al., 1996).

It is useful to further distinguish the possible climatic mechanisms discussed above in terms of the fundamental nature of the underlying dynamics. This is not always a straightforward task, as the distinction between self-sustained unstable oscillations and stochastically forced stable oscillations based on classical diagnostics may not be obvious (Saltzman et al., 1981). Self-sustained nonlinear oscillations result from a phase-space bifurcation of the system's dynamics (e.g., Hopf bifurcation—see Quon and Ghil, 1995; Chen and Ghil, 1995; 1996). Furthermore, such self-sustained nonlinear oscillatory behavior may exhibit a dependence on the external control parameters of the nonlinear system, and frequency modulation is also possible (e.g., Tziperman et al., 1994;

Jin *et al.*, 1994) if the phase-space character of the system undergoes lower frequency changes. Such oscillatory behavior tends to exhibit chaotic intermittent oscillations (see Lorenz, 1990) and furthermore, is assumed to be obscured by the noise present in the climate system. In contrast, stochastically excited, damped stable oscillations can arise in both linear and nonlinear systems. Such oscillations arise from the excitation of the natural eigenmodes of a stable system by stochastic colored-noise forcing (e.g., Hasselmann, 1988; Mysak *et al.*, 1993; Weaver and Sarachik, 1991; Schmidt and Mysak, 1996; Delworth *et al.*, 1993; Latif and Barnett, 1994). Any means of exploratory signal detection in climate studies should be sufficiently general to identify, though perhaps not distinguish, stochastically excited or self-sustained oscillatory behavior, since neither can a priori be eliminated based on theoretical or dynamical considerations.

2. Traditional Methods of Oscillatory Climate Signal Detection

A variety of techniques have been applied to the problem of signal detection in observational and dynamical model-generated climate data. Such techniques have typically employed univariate methods for isolating narrowband peaks in the power spectrum of climate time series based on spectral estimation methods such as Blackman–Tukey or maximum entropy spectrum analysis (e.g., Brillinger, 1981; Marple, 1987). Traditional attempts to exploit the mutual information available in spatially distributed climate records have involved principal component analysis (PCA) or related orthogonal multivariate spatiotemporal decompositions (Preisendorfer, 1988; Bretherton *et al.*, 1992) followed by spectral analysis of the time series of the resultant spatial modes (e.g., Trenberth and Shin, 1984; Deser and Blackmon, 1993; Tanimoto *et al.*, 1993). Only recently have methods been developed [e.g., principal oscillation patterns (POPs)—Hasselmann, 1988; multichannel singular spectrum analysis (M-SSA)—Keppenne and Ghil, 1993; and multitaper frequency-domain singular value decomposition (MTM-SVD)—Mann and Park, 1994] which simultaneously exploit both the coherent spatial structure and the narrowband frequency-domain structure of climatic signals for more efficient spatiotemporal signal detection. Furthermore, the properties of climatic noise and proper null hypothesis testing in the context of multivariate signal-detection approaches have only recently begun to receive proper attention (e.g., Mann and Park, 1996b; Allen and Robertson, 1996).

In this section, we motivate a particular model—that of spatiotemporal colored noise—as the null hypothesis for climate variability. A spatiotemporal model is invoked because of the intrinsic large-scale spatial structure of climatic variations. A "colored-noise" model is invoked because the complicated dynamics of the climate system lead to stochastic variations with a frequency-domain structure more complicated than simple (e.g., Gaussian white-noise) models. This model must be sufficiently well rejected if we are to infer the existence of (i.e., "detect") a signal in a climatic data field. We argue that such climate signals should be associated with patterns that exhibit wider spatial coherence than the underlying noise, with narrowband frequency-domain signatures. Under such assumptions for climatic data, we show how traditional methods for signal detection suffer from a number of weaknesses. We motivate instead the MTM-SVD methodology, which overcomes many such weaknesses and provides certain optimal features in multivariate signal detection and reconstruction. We demonstrate that the MTM-SVD method produces correct inferences when applied to known (i.e., specified synthetic) spatiotemporal colored-noise processes, focusing on spatially correlated "red" noise (including that which is considerably "redder" than estimates for actual climate data). When applied to synthetic data examples, MTM-SVD provides excellent fidelity in signal detection and reconstruction. Finally, we demonstrate that the approach does not suffer significantly when substantial temporal and spatial inhomogeneities, typical in real climate data, are introduced into the synthetic example.

2.1. Signal and Noise Assumptions: A Synthetic Data Set

We introduce here a synthetic example with the basic signal and noise attributes inferred for observed climate data. The synthetic data are constructed on a grid resembling a cartesian projection of the spherical globe such as is typically used for gridding actual spatial climate data (Fig. 1). The gridded network has uniform monthly sampling and a duration $N = 1200$ months (100 years).

The data set is constructed so as to contain two irregular oscillatory signals and a trend, each widely correlated over the synthetic global domain, linearly added to noise which exhibits near-neighbor spatial correlation and an underlying red-noise temporal autocorrelation structure. Thus constructed, the noise, while spatially correlated, does not exhibit the large-scale coherent structure associated with the low-frequency climate "signals." This latter distinction, motivated for both theoretical and observational-based considerations described below, underlies the reason that multivariate analysis can greatly enhance the effective signal-to-noise ratio and efficiency of signal detection and reconstruction.

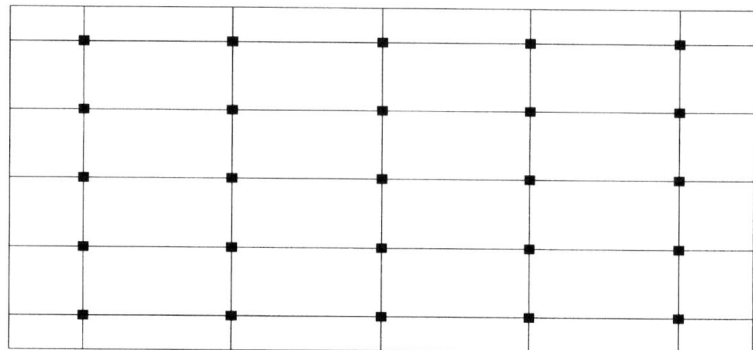

FIG. 1. Global cartesian grid showing spatial sampling of synthetic data network. Sampling of 25 gridpoints is equally distributed in both longitude (72° separation between gridpoints) and latitude (36° separation between gridpoints), with data centered at the "equator" and in "subtropical" and "subpolar" zonal bands in the two "hemispheres" of the synthetic domain. [From Mann (1998).]

Noise Component

We adopt a model of spatially correlated colored noise that is motivated by theoretical models for stochastic climate variability. In the absence of any complex dynamics, the inertia of the ocean and other slow-response components of the climate system alone tend to integrate any high-frequency (often approximated as "white") noise forcing provided synoptic-scale "weather" forcing (see Hasselmann, 1976), altering the temporal characteristics of the noise but preserving the limited, near-term spatial correlation structure of the noise.

The simplest mathematical description of such an integrating noise process in the context of discretely measured variables such as monthly mean climate data, is the first-order autoregressive (AR(1)) red-noise process (see Gilman *et al.*, 1963), specified by the statistical model

$$y_t = \rho y_{t-1} + w_t \tag{1}$$

(where w_t is a white-noise "innovation" sequence, with variance σ^2) and characterized by the power spectrum

$$S(f) = S_0 \frac{1 - \rho^2}{1 - 2\rho \cos(f/f_N) + \rho^2}, \tag{2}$$

where the average power S_0 is related to the white-noise variance,

$$S_0 = \sigma^2/(1 - \rho^2). \tag{3}$$

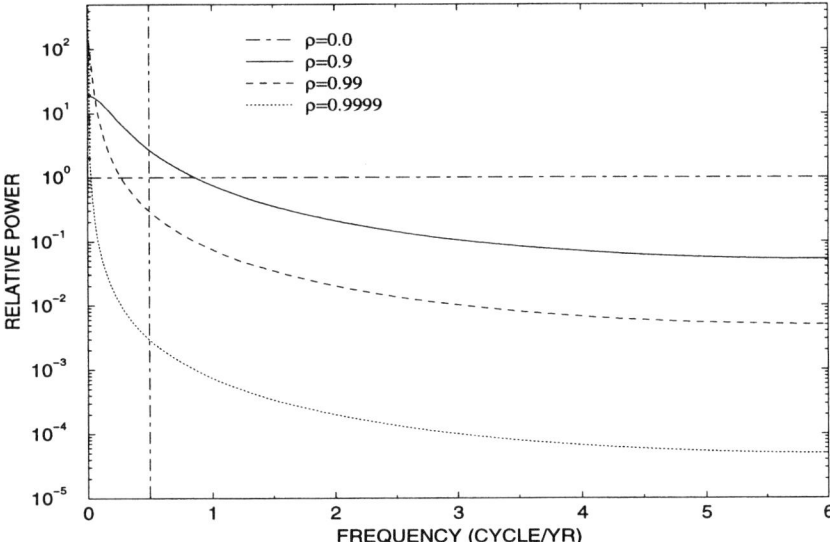

FIG. 2. Power spectrum of ideal monthly sampled AR(1) red-noise processes with varying levels of month-to-month autocorrelation ρ. The vertical line separates off the "interannual" ($f < 0.5$ cycle/yr; periods greater than $T = 2$ year) regime of the spectrum. [From Mann (1998).]

Figure 2 shows the power spectrum for ideal monthly sampled AR(1) red-noise processes of increasing levels of autocorrelation. In the AR(1) red-noise model, autocorrelation decays exponentially so that the decorrelation time scale of the noise τ is related to the lag-1 autocorrelation coefficient ρ by $\tau = -\Delta t/\log \rho$, where Δt is the temporal sampling interval. ρ (and less directly τ) is a parameter that measures the relative "smoothness" of the noise spectrum, i.e., how rapidly the amplitude of the noise spectrum varies with frequency. This smoothness can be quantified in terms of the relative gain over the bandwidth of the spectrum estimator, which we will define by the factor

$$F(f) = \Delta f_{BW} \frac{d \log S(f)}{df}, \qquad (4)$$

where a spectrum bandwidth $\Delta f_{BW} \approx 0.04$ cycle/year is typically applicable in studies of multidecadal-to-centuries-duration climate data sets (see Section 2.1). Thus defined, F measures the relative "inverse" of smoothness. A smoothly varying noise spectrum can be characterized by the local or global (in the frequency domain) fulfillment of the condition that F be small compared to unity (i.e., $1/F$ large). For $F \geq 1$, we might expect the "smoothness" assumption to begin to break down, and for $F \gg 1$, we certainly do.

Given the typical bandwidth indicated above, we can consider the estimated smoothness for the different cases described in Figure 2. For $\rho = 0$, we observe a uniform "perfect" inverse smoothness factor $F = 0$ for all frequencies. For the "moderate" case $\rho = 0.9$, we have a maximum value of the parameter $F_{MAX} \approx 0.11$ near $f = 0.2$ cycle/year, and an average value over the interannual frequency range of interest $f < 0.5$ cycle/year, of $F_{AVE} \approx 0.09$. For the more strongly red case $\rho = 0.99$, the corresponding values are $F_{MAX} = 0.8$ (near $f = 0.02$ cycle/year) and $F_{AVE} = 0.2$. Finally, for the nearly singular case, we have $F_{MAX} = 8$ (near $f = 0$) and $F_{AVE} = 0.6$. Roughly speaking, then, we might consider the noise spectrum as smoothly varying throughout the interannual frequency range for the case $\rho < 0.9$. For the case $0.9 \lesssim \rho < 0.99$, we can consider the noise spectrum as varying smoothly over most of the interannual frequency interval, though perhaps less so near zero frequency. As $\rho \to 1$, the AR(1) red-noise process approaches a pure random walk $y_t = y_{t-1} + w_t$, characterized by first-order nonstationarity, an infinite decorrelation time scale τ, and a spectrum that is singular at $f = 0$. As the random-walk condition is approached, the noise spectrum may violate our smoothness assumption. As shown in Section 2.1, the best-fit red-noise spectrum for instrumental climate data falls in the white ($\rho = 0$) to moderately red ($\rho = 0.9$ for monthly data—i.e., $\tau \approx 0.8$ years) range and should satisfy the requirements of a smoothly varying noise background.

AR(1) red noise accurately describes the physical model of natural variability for a simple stochastically forced energy balance of the climate which contains a mixed-layer ocean. It can also be shown to be the limiting case of more complex stochastic models of climate which allow for exchange of heat with the deep ocean (Wigley and Raper, 1990). The latter noise model in general requires the specification of a number of poorly constrained physical parameters, and is statistically described by higher-order AR models. A combination of dynamical considerations and parsimony thus might tend to favor the AR(1) red-noise model. Indeed, empirical studies of a wide variety of proxy and instrumental climate data (Gilman *et al.*, 1963; Kutzbach and Bryson, 1974; Allen and Smith, 1994; Mann and Lees, 1996) suggest that the AR(1) red-noise model provides an excellent description of the background climate noise spectrum. Nonetheless, the ideal null hypothesis accommodates both simple AR(1) red noise and more general colored-noise processes. In Section 3.3 we introduce a means for employing such a more general colored-noise null hypothesis in signal detection. Nonetheless, for demonstrative purposes, we here consider the null hypothesis of climatic noise modeled as having the temporal correlation structure described by the AR(1) red-noise model, and short-range spatial correlation structure. Typical estimates of the temporal decorrelation time scales for monthly gridded surface temperature data, for example, are $\tau \lesssim 1$ year (Allen and Smith, 1994; Mann and Lees, 1996). Estimates of temporal decorrelation scales in actual observational data are discussed in more detail in Section 2.1. While the spatial decorrelation length scale d tends to vary somewhat with season (Livezey and Chen, 1983; Briffa and Jones, 1993), estimates from both model-based (Madden *et al.*, 1993) and observational (Kim and North, 1991; Mann and Park, 1993) data indicate an approximate value of $d = 1500-2000$ km for monthly surface temperature data. In keeping with the above qualitative description of climatic noise, we prescribe a spatiotemporal AR(1) red-noise background with a roughly $d = 1.5$-grid spacing decorrelation length scale, and a temporal decorrelation time scale of $\tau \approx 0.9$ year ($\rho = 0.9$) in the synthetic monthly data set.

Signal Component

Typical climate signals (e.g., the El Niño/Southern Oscillation or ENSO) tend to be associated with large-scale (i.e., global or hemispheric-wide) perturbations of the coupled ocean–atmosphere system. Such signals are detectable not only in climatic measurements in the regions where the intrinsic climate dynamics are important (e.g., the tropical Pacific in the case of ENSO—see Cane *et al.*, 1986); through their altering effect on planetary wave propagation and global atmospheric circulation patterns, they lead to substantial perturbations in remote regions (e.g., Horel and Wallace, 1981). Such signals are thus detectable in part because of their hemispheric- or global-scale spatial organization. The patterns of expression of ENSO in surface temperature (Ropelewski and Halpert, 1987) and precipitation (Halpert and Ropelewski, 1992) are clearly global in extent and have been theoretically shown to be consistent with the influence of tropical heating anomalies on the planetary wave structure of the extratropical atmosphere (see, e.g., Horel and Wallace, 1981). There is recent evidence, both in observational studies (e.g., Dettinger *et al.*, 1995; Ghil and Vautard, 1991; Mann and Park, 1993; 1994; 1996b; Schlesinger and Ramankutty, 1994; Mann *et al.*, 1995b) and in coupled ocean–atmosphere model simulation studies (e.g., Latif and Barnett, 1994; Delworth *et al.*, 1993), for oscillatory climate mechanisms with similar global-scale influence at decadal and longer time scales.

Most theoretical models describe such signals as having a quasi-oscillatory character. Positive and negative feedbacks, and delayed oscillator coupled mechanisms, can allow for oscillatory behavior in either a linear or a nonlinear dynamical context. Either intrinsic nonlinearities or stochastic forcing can modulate both amplitude and phase, leading to finite spells, or episodes, of coherent oscillatory behavior. Furthermore, frequency modulation (e.g., in the case of ENSO—see Tziperman et al., 1994; Jin et al., 1994) can result from changes in external governing parameters. A proper statistical model for oscillatory climate signals must thus describe a narrowband but not strictly periodic mode of variability with spatial scale structure that is coherent (though perhaps quite variable in sign or phase) at large spatial scales. Climatic trends that are inconsistent with the noise null hypothesis can be treated as oscillatory signals with zero frequency (i.e., infinite period). Other types of climate signals (e.g., volcanic climate perturbations) may have truly event-like character that is best described by alternative statistical models (e.g., wavelet-based generalizations of the frequency-domain methods discussed below—see Lilly and Park, 1995; Park and Mann, 1999).

The synthetic data set exhibits the key features of our conceptual model of the climate system. Slowly modulated quasi-oscillatory low-frequency components and a secular trend are superposed on a spatially and temporally autocorrelated noise component, with the relative importance of each varying by location. We construct three synthetic signals that exhibit the kinds of complexity (e.g., amplitude, phase, and frequency modulation) that we might expect to encounter in true climate signals. The first signal is a secular trend with a half-period cosine shape describing a variable amplitude "warming" trend in most locations. Some locations exhibit the opposite sign or a vanishing amplitude. This signal represents an analog for a spatially variable global warming signal. The second signal represents an interdecadal oscillation with phase/amplitude modulation that vanishes in a global average due to phase cancellation over the domain. The third signal exhibits the most complex characteristics, with uniform amplitude, but partial phase cancellation, an amplitude trend with periodic modulation and linear ramp, and frequency modulation with a rapid transition between 3- and 5-year periodicity during the middle 40 years. This signal exhibits a poleward propagating phase pattern.

The amplitudes of the signals are prescribed so that the total signal variance is equal to the total noise variance (i.e., the aggregate signal-to-noise variance ratio is unity). The secular trend describes 56% of the raw data variance, the interdecadal signal 25%, and the interannual signal 8%. The residual 43% variance is explained by the spatially correlated red noise. Note the similarity between this imposed breakdown of variance and the empirical signal/noise decompositions of Mann and Park (1994; 1996b), recounted in Sections 4.1 and 4.2. In these data sets the identified signals

TABLE I Description of the Three Synthetic Examples and Noise in the Synthetic Example, Indicating the Spatial and Temporal Characteristics of Each Signal (Spatial Phase and Amplitude Pattern, and Pattern of Temporal Modulation), Signal Period (or Period Range) in Years, Frequency (or Frequency Range) in Cycles / Year, and Maximum Regional Peak Amplitude [From Mann (1998).]

Signal	Spatial char	Temporal char	T (years)	f (cycle/yr)	Max amp.
Trend	Variable amp./ sign	Half-cosine trend	200	0.005	1.0
Interdecadal oscillation	Variable amp./ phase	Amp. mod	15	0.065	1.0
Interannual oscillation	Uniform amp./ variable phase	Amp./freq. mod	3–5	0.33–0.2	1.0
Red noise	Near-neighbor spatial correlation	AR(1) red noise			1.0

consume a somewhat smaller proportion of the data variance, roughly 40%. The residual 60% variance was attributed to the colored-noise background. The characteristics of the signals and noise are summarized in Table I. In Figure 3, we show the spatial and temporal patterns of the three signals described above, while in Figure 4, we show the time reconstruction for a reference site (center gridpoint). The relative spatial pattern is depicted by a vector map in which the angle represents the relative phase and the length indicates the relative amplitude of the signal at each gridpoint.

Comparison with Actual Climate Data

In this section, we estimate the signal and noise properties of actual instrumental climate data to help motivate our assumptions regarding signal and noise in the preceding sections. We make use of the historical gridded temperature data used by Mann and Park (1994) in their multivariate analysis of global temperature variations. In Figure 5, we show the spectra (as estimated by the multitaper method—see Section 2.2) along with the best-fit red-noise background for a few instrumental gridpoint temperature series in the subset of 449 nearly continuous 100-year gridpoint temperature records over the globe (see Figure 16).

In almost every one of the 449 gridpoint series, the null hypothesis of white noise (i.e., AR(1) noise with $\rho = 0$) is rejected at a very high level of likelihood, with the best-fit values of ρ ranging from 0.09 to 0.80 and averaging $\rho \approx 0.35$. The null hypothesis of red noise is only *weakly*

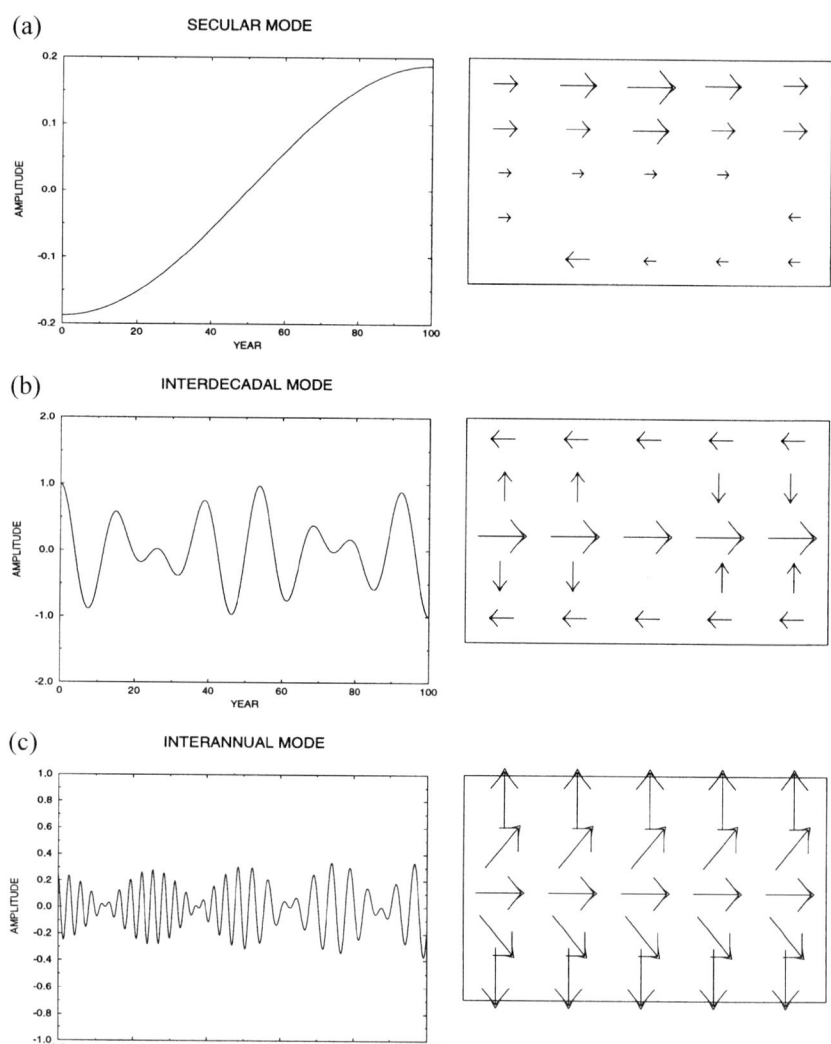

FIG. 3. Temporal (left) and spatial (right) patterns of synthetic signals showing (a) secular mode, (b) interdecadal mode, and (c) interannual mode. Conventions are described in the text. [From Mann (1998).]

TIME SERIES FOR REFERENCE GRIDPOINT

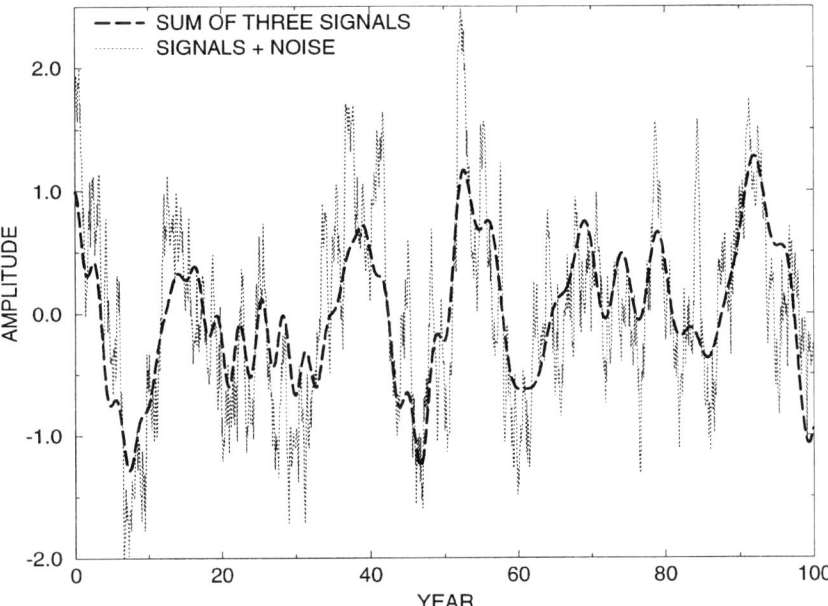

FIG. 4. Time signal and signal + noise for reference (central) gridpoint time series. [From Mann (1998).]

rejected, however. For example, due to chance coincidence alone, we would expect 30 peaks to randomly exceed the 90% confidence level over the positive Nyquist interval ($f = 0$ to $f = 6.0$ cycle/year) for a realization of a true AR(1) red-noise process with the bandwidth $NW = 2$ employed in the spectrum estimation (see Section 2.2). In contrast, the typical temperature gridpoint yields 35–45 peaks that exceed that level. This small discrepancy between expected rates of false detection and observed rates of signal detection implies the existence of additional structure in the climate spectrum which is not consistent with red noise. We hypothesize that such additional structure implies the existence of a small number of distinct band-limited processes superposed on a stochastic red-noise background. We thus argue that (a) moderate ($\rho < 0.9$) red noise provides an excellent null hypothesis for the noise background, and (b) there is evidence for a small number of signals in addition to the noise background. The foregoing analysis cannot establish whether the latter "sig-

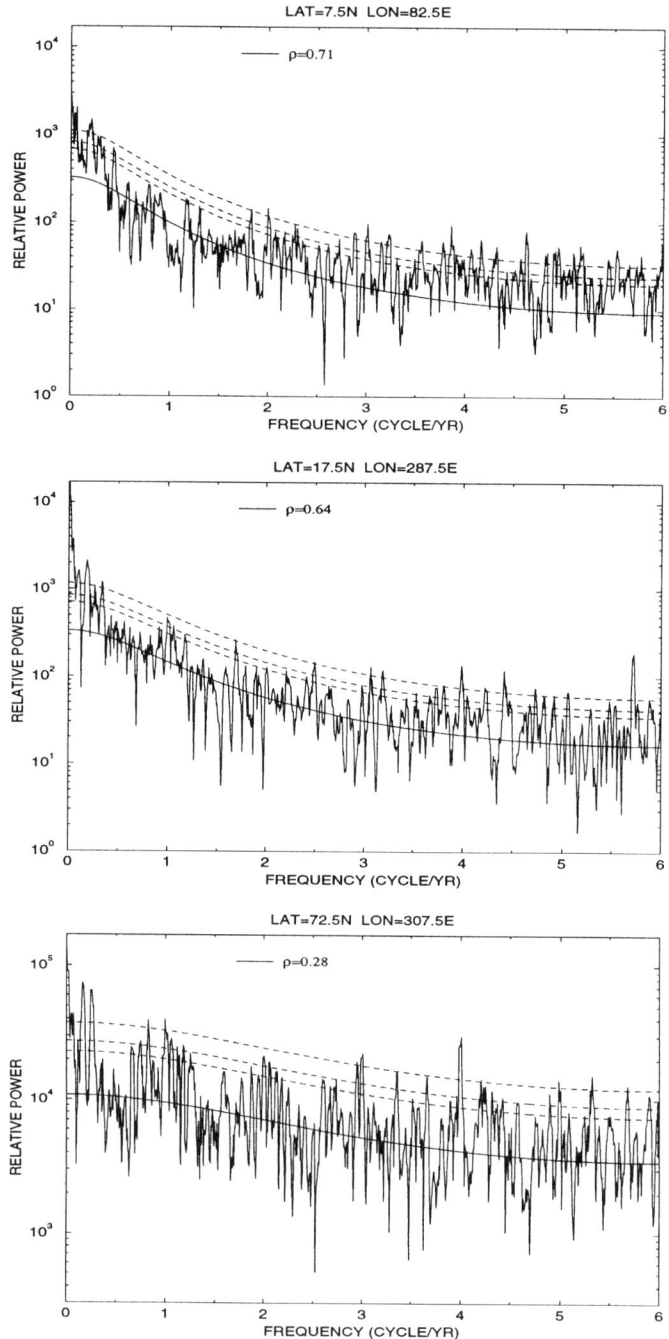

nals" do indeed represent spatiotemporally consistent signals in the multivariate data, and is limited by its assumptions of a strict AR(1) noise model. We introduce in Section 3.1 a methodology for signal detection that does not suffer either of these limitations. First, however, we review the traditional approaches to oscillatory climate signal detection.

2.2. Conventional Approaches to Signal Detection

Univariate Signal Detection

While a variety of traditional spectral analysis methods (e.g., Blackman–Tukey) have been widely employed in the analysis of geophysical processes (see, e.g., the review by Brillinger, 1981 and references therein), specialized methods have more recently been developed that are more faithful in their underlying assumptions to the irregular oscillatory behavior expected of climatic signals. Among such methods are multitaper spectral analysis (Thomson, 1982; Park *et al.*, 1987; Percival and Walden, 1993), which employs multiple orthogonal data tapers to describe phase- and amplitude-modulated structures, and singular spectrum analysis (Vautard and Ghil, 1989; Ghil and Vautard, 1991; Ghil and Yiou, 1996), which makes use of anharmonic basis functions derived from the lagged covariances of the data series.

These univariate spectral analysis approaches have been used to detect and reconstruct the complicated signals present in climate data (MTM—see Thomson, 1990; Kuo *et al.*, 1990; Park and Maasch, 1993; Mann and Park, 1993; Thomson, 1995; SSA—see Ghil and Vautard, 1991; Yiou *et al.*, 1991, 1993; Allen and Smith, 1994; Schlesinger and Ramankutty, 1994; Lall and Mann, 1995). Furthermore, considerable attention has been paid to assure proper null hypothesis testing in climate studies for both SSA (e.g., Allen and Smith, 1994) and MTM (Mann and Lees, 1996). Rather than focusing on a comparison of these methods (see, e.g., Thomson, 1982; Ghil and Yiou, 1996), we here focus on the application of the MTM method to univariate signal detection alone, postponing any intermethod comparison to the discussion of multivariate signal-detection techniques and the multivariate generalization of MTM described in Section 3.

FIG. 5. Multitaper spectra of three different 100-year-long monthly land air and sea surface temperature gridpoint records over the globe based on time-frequency bandwidth factor $NW = 2$ and $K = 3$ tapers, along with robustly estimated median red-noise level and 90, 95, and 99% confidence limits for significance relative to red noise—see Section 2.2. [From Mann (1998).]

In the multitaper method, one determines for a given time series, $\{x\}_{n=1}^{N}$, a set of K orthogonal data tapers and K associated tapered Fourier transforms or "eigenspectra":

$$Y_k(f) = \sum_{n=1}^{N} w_n^{(k)} x_n e^{i2\pi f n \Delta t}, \qquad (5)$$

where $\Delta t = 1$ month is the sampling interval and $\{w_n^{(k)}\}_{n=1}^{N}$ is the kth member in an orthogonal sequence of Slepian tapers, $k = 1, \ldots, K$. The "time-frequency bandwidth parameter" defined by $NW = p$ defines a particular family of eigentapers. Only the first $K = 2p - 1$ tapers are usefully resistant to spectral leakage, so that the choice of K and p represent a trade-off between spectral resolution and the degrees of freedom (which can be used to constrain the variance of the spectral estimators). In the context of climate studies of roughly century duration, $NW \equiv p = 2$ and $K = 3$ provide a good compromise (Mann and Park, 1994; Mann and Lees, 1996) between the resolution appropriate to resolve the natural bandwidths of climatic signals and the stability of spectral estimates. The set of K tapered eigenspectra have energy concentrated within a bandwidth of $\pm p f_R$ centered on a given frequency f, where $f_R = (N\Delta t)^{-1}$ is the Rayleigh frequency. Thus, the choice $p = 2$ provides a full bandwidth of spectral estimation $\Delta f_{BW} = 2p f_R \approx 0.04$ cycle/year for a 100-year data series. Each of the K eigenspectra represents statistically independent local averages of the spectral information near f, under the assumption of a smoothly varying colored ("locally white") spectral background. As explained in Section 2.1, this assumption holds up very well for actual climate data which exhibit a weak-to-moderate red-noise background. Figure 6 shows the three orthogonal data tapers for the case $K = 3$, along with a sinusoid modulated by each of the K eigentapers. From the latter plot, it is evident that the multitaper analysis can provide a description of an irregular narrowband oscillatory signal centered at a particular frequency f through the variety of amplitude and phase modulations that can be described by a suitable linear combination of K independently tapered carrier oscillations. Each of the K spectral degrees of freedom available for each time series at a given frequency f will provide statistical information in the multivariate extension described in Section 3. In univariate applications, these independent estimates are combined through a weighted average of the eigenspectral estimates to provide a spectral estimate with optimal spectral resolution/variance trade-off properties (Thomson, 1982),

$$s(f) = \frac{\sum_{k=1}^{K} \lambda_k |Y_k(f)|^2}{\sum_{k=1}^{K} \lambda_k}. \qquad (6)$$

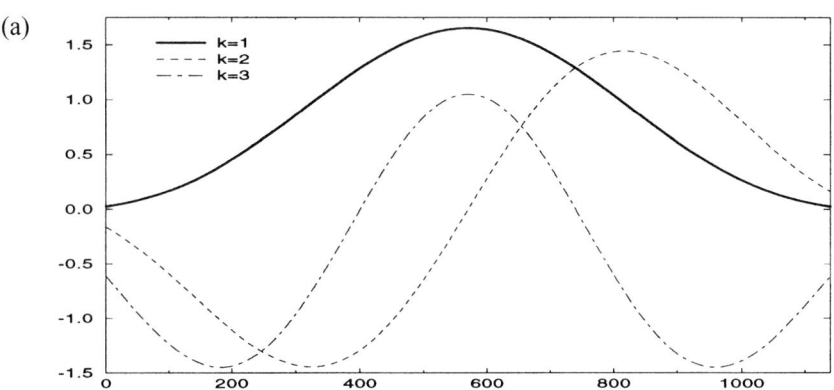

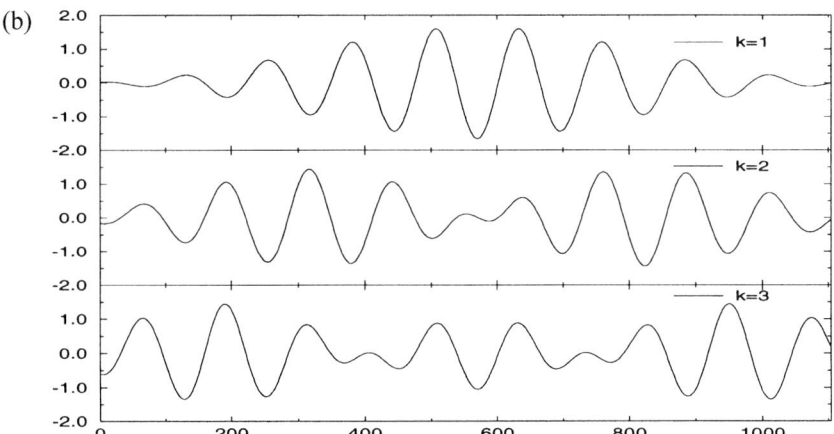

FIG. 6. (a) The first three orthogonal Slepian data tapers for the case $K = 3$, $NW = 2$, and (b) the components of a fixed-amplitude phase-coherent sinusoidal oscillation described by the modulating envelope associated with each of the $K = 3$ data tapers. [From Mann (1998).]

An "adaptively weighted" estimate of the spectrum can be calculated as

$$s(f) = \frac{\sum_{k=1}^{K} b_k^2(f) \lambda_k |Y_k(f)|^2}{\sum_{k=1}^{K} b_k^2(f) \lambda_k}, \qquad (7)$$

where b_k is a data-adaptively determined weighting function of the eigenspectra that seeks to minimize broadband leakage in the spectrum (Thomson, 1982).

We here show the results of univariate MTM spectral analysis applied to the problem of detecting signals in the synthetic data set (Fig. 7). We use the procedure of Mann and Lees (1996) to provide robust estimates of the estimated red-noise background and significances of narrowband peaks. Since the phase of each oscillation varies over the spatial grid, there is some cancellation of the oscillatory components in the average across series. Note that the MTM spectra of the "global average" do not detect the interdecadal signal, constructed to vanish in a global average, as significant. Partial phase-cancellation of other signals also diminishes the usefulness of large-scale spatial averaging. Although interdecadal and interannual peaks are detected for the "reference" gridpoint, the secular trend, small at that gridpoint, is not recognized as significant at the 95% level. The secular trend and interdecadal peak are clearly detected in the spectrum for the "northwest" grid point, but it is difficult to identify any consistent interannual peaks in this spectrum or that of the "reference" gridpoint, and there are several spurious peaks (sampling fluctuations from the noise background) that rival the true signals in their prominence. Thus, on one hand, large-scale spatial averaging is often an ineffective means of signal/noise ratio enhancement. On the other hand, signal-to-noise ratios in the univariate "regional" signal-detection approach are too low for consistent detection of large-scale signals. It is thus clear that the mutual information available in the spatially distributed data must be used in a more sophisticated way for effective spatiotemporal signal detection. This is particularly true in exploratory analysis where the spatial structures at different frequencies are not known a priori.

PCA + Spectral Analysis

A common approach to spatiotemporal signal detection in geophysical applications is based on some variant of principal component analysis (PCA), in which a singular-value decomposition (SVD) is performed on the

---

FIG. 7. Multitaper spectra of time series from (a) global average over domain, (b) central gridpoint, and (c) extreme northwest gridpoint, along with robustly estimated median red-noise level and 90, 95, and 99% confidence limits for significance relative to red noise. [From Mann (1998).]

(a)
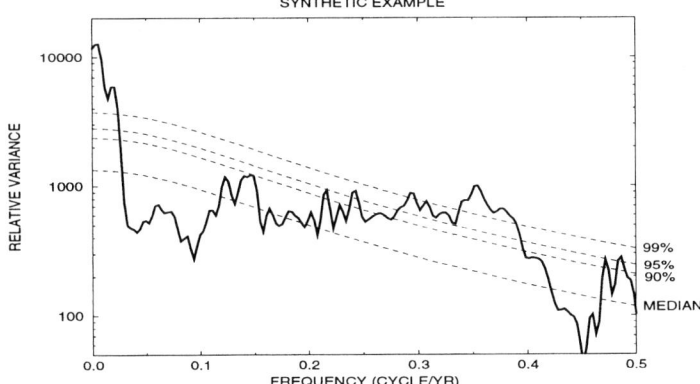

(b)
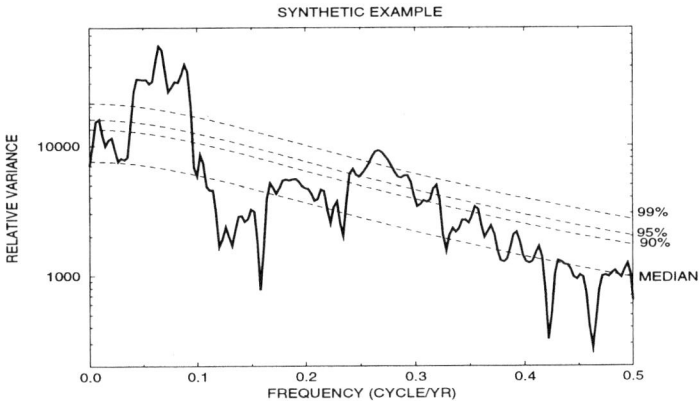

(c)
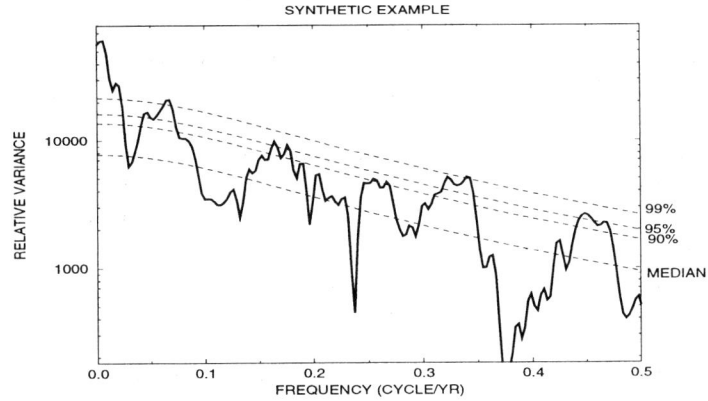

data matrix, followed by spectral analysis (PCA + SA) of the time series of the independent spatial modes (Trenberth and Shin, 1984; Deser and Blackmon, 1993; Tanimoto et al., 1993; Venegas et al., 1996).

Consider a set of individual data series $x^{(m)}$ ($m = 1, 2, \ldots, M$) of length N time units (e.g., months or years) centered to represent departures from the respective long-term means. Typically, each series is normalized by its standard deviation. The resulting demeaned and normalized series are termed "standardized" series.

The standardized spatiotemporal data can then be written as a data matrix:

$$X = \begin{bmatrix} w_1 x_{t_1}^{(1)} & w_2 x_{t_1}^{(2)} & \cdots & w_M x_{t_1}^{(M)} \\ w_1 x_{t_2}^{(1)} & w_2 x_{t_2}^{(2)} & \cdots & w_M x_{t_2}^{(M)} \\ \vdots & & & \\ w_1 x_{t_N}^{(1)} & w_2 x_{t_N}^{(2)} & \cdots & w_M x_{t_N}^{(M)} \end{bmatrix}, \qquad (8)$$

where $t_1, t_2, \ldots, t_N$ spans over the N time samples, and $m = 1, 2, \ldots, M$ spans the M (e.g., individual gridpoint) different series. w_m might, for example, indicate weightings by gridpoint area.

The data matrix is decomposed by singular-value decomposition,

$$X = \sum_{k=1}^{M} \lambda_m \mathbf{u_m}^\dagger \mathbf{v_m}, \qquad (9)$$

into its dominant spatiotemporal eigenvectors, where the M-vector or empirical orthogonal function (EOF) $\mathbf{v}_m$ describes the relative spatial pattern of the mth eigenvector, the N-vector $\mathbf{u}_m$ or principal component (PC) describes its variation over time, and the eigenvalue (the square of the singular value) λ_m is the associated fraction of described data variance. The dagger on the vector $\mathbf{u_m}$ indicates the conjugate transpose.

We demonstrate the application of PCA to the synthetic data set described in Section 2.1. Four eigenvalues (each indicating the fractional data variance explained by an associated empirical eigenvector which describes the temporal variation of a particular fixed anomaly pattern) are established as significant relative to spatially correlated noise in the multivariate data set (Fig. 8). The significance criterion is based on the PCA selection "rule N," modified to take into account the reduction in spatial degrees of freedom owing to spatial autocorrelation (Preisendorfer, 1988). Estimating a spatial decorrelation scale of $d = 1.5$ gridpoints leads to $N' \approx N/d = 11.7$, so that only eigenvectors with $\lambda > 1/N' = 0.085$ are to be retained. Furthermore, there is a clear break in the eigenvalue

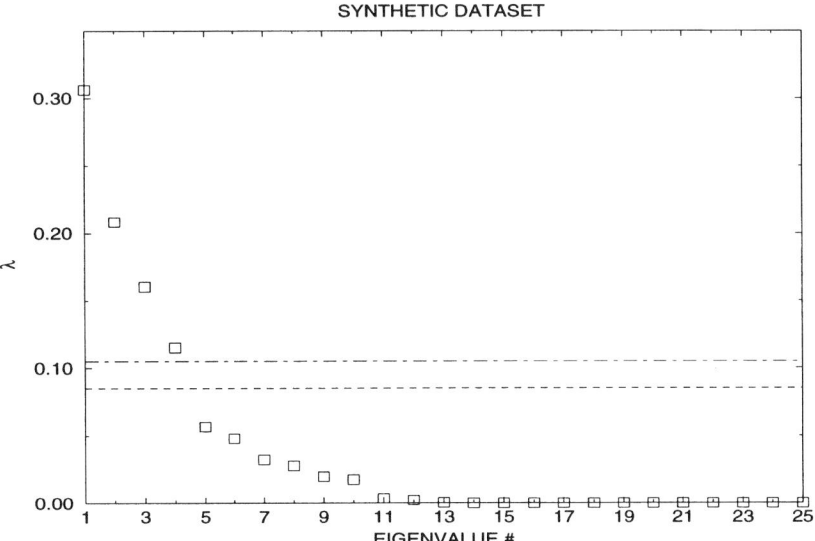

FIG. 8. Eigenvalue spectrum of the PCA decomposition of the synthetic data set. Four eigenvectors are indicated as being statistically significant taking into account spatial autocorrelation of the gridpoint data. Thresholds for significance are shown based on the two different calculations described in the text. [From Mann (1998).]

spectrum from its red-noise floor between eigenvalues 4 and 5, so the selection of four eigenvectors seems quite natural in this case.

The decomposition provided by PCA exhibits several clear shortcomings. An immediate problem is that the PCA procedure detects four significant statistically independent modes of variation in the data when we know a priori that only three modes of variation are distinct from the red-noise background.

Furthermore, the power spectra of the statistically significant PCs (Fig. 9) present a muddled picture of signal and noise in the data set. PC #1 describes a pattern of variability which exhibits dual dominant time scales including a significant trend and significant narrowband variance in the 3 to 5-year interannual range. The reader will note a striking similarity to the spectrum of the globally averaged data shown earlier in Figure 7a. PC #1, to a very good degree of approximation, describes the globally in-phase mode of variation in the data set. This component is slightly different from the global mean because variations that are 180° out of phase project oppositely onto the global mean. Thus, the principal mode of

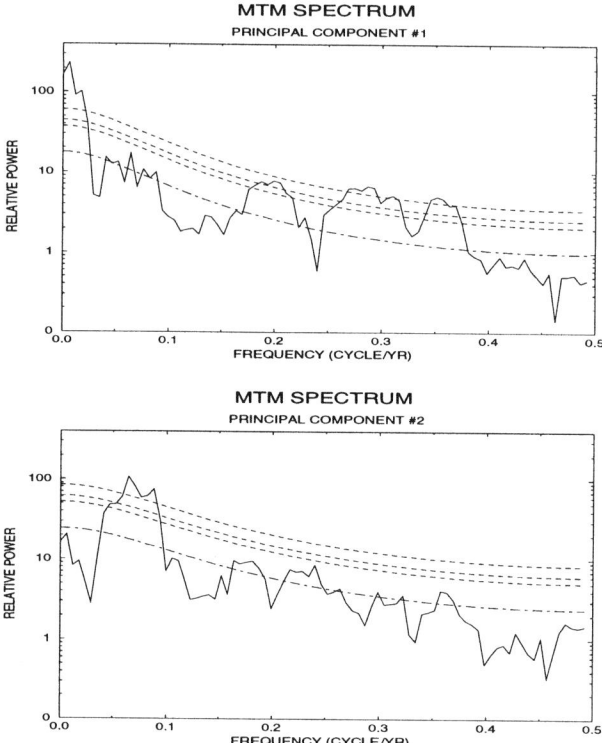

FIG. 9. Multitaper spectra of the significant principal components (1–4) along with significance levels relative to red noise. [From Mann (1998).]

the PCA has no simple "physical" interpretation, representing a combination of incomplete projections of two of the signals—secular trend and interannual signal—which project onto the global mean. PCs #2, 3, and 4 describe various combinations of the residual, spatially heterogeneous component of the multivariate data. The modulated interdecadal oscillation appears as a peak of varying prominence in each of these three PCs. The interannual signal is scattered in varying degree among each of the PCs. The noise background, furthermore, is not consistently decomposed among the four PCs, with PC #3 exhibiting a considerably whiter noise background than the others. The misidentification of signal and noise [arises here] from fundamental weaknesses in the PCA + SA signal-detection approach. The primary weakness results from the performance of two consecutive statistical operations which have conflicting optimality properties. PCA performs a time-optimal variance decomposition through

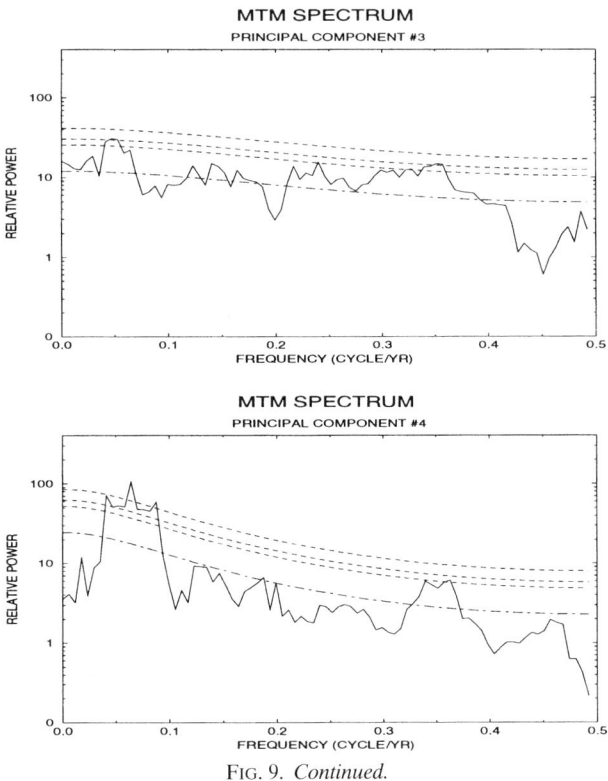

FIG. 9. *Continued.*

a Karhunen-Loeve expansion of the data set in the time domain, appropriate for a random or broadband multivariate process. However, as discussed earlier, there is considerable evidence for narrowband processes in observed climate data with characteristics similar to those we have imposed in our synthetic example. When such narrowband frequency-domain structure is present, the thoroughly unoptimal frequency-domain properties of the time-domain decomposition become apparent (see the discussion by Brillinger, 1981, Chapter 9). In contrast, a frequency-domain Karhunen-Loeve expansion provides an optimal decomposition of the data variance for this latter case (Thomson, 1982). The combination of a red-noise background and narrowband multivariate processes in the synthetic data set thus cannot be efficiently separated in the PCA + SA approach. Furthermore, because there is no phase information (see, however, the "complex" PCA method described below) in the PCA decomposition, the propagating phase structure in the signals cannot be correctly described by the empirical eigenvectors. Rather, the eigendecomposition must artificially describe such phase information in terms of multiple standing waves.

Nonetheless, a variety of generalizations of PCA have been developed which attempt to ameliorate several of the problems noted above, through various modifications or alternative spatiotemporal variance decompositions. Below we discuss such methods, pointing out the relative strengths and weaknesses of each approach, and emphasizing those particular weaknesses or limitations which are overcome by the MTM-SVD approach described in Section 3.

Multivariate AR

While typically applied to the problem of spatiotemporal interpolation of data fields in climate studies (e.g., Wikle and Cressie, 1996; Kaplan *et al*., 1997), Markovian (i.e., AR(1) or higher-order AR) spatiotemporal models do also provide a means of multivariate spectrum estimation and signal detection (see Marple, 1987). Such multivariate AR methods of spectral analysis offer the drawbacks, however, that (a) they assume strict stationarity of the data and (b) they provide less than optimal resolution/variance trade-off properties in spectral estimation.

Principal Oscillation Patterns

Principal oscillation patterns or POPs (Hasselmann, 1988; Penland, 1989; Xu, 1993; Von Storch *et al*., 1995) exploit Markovian structure in the data in a dynamical context distinct from that of the conventional multivariate AR approach. POPs offer a philosophical appeal under certain assumptions for the governing dynamics; they invoke a specific dynamical model—stochastically forced damped linear oscillatory behavior—in multivariate spectral estimation. Furthermore, POPs are readily generalized to incorporate spatially variable phase information (complex POPs or CPOPs—see Bursor (1993)). As long as the underlying model of stable linear dynamics is appropriate, and the climatic data series to be analyzed are long, POPs or CPOPs provide a useful means of signal detection. On the other hand, the specificity of the subsumed inverse model limits the usefulness of POPs as an exploratory data analysis tool, when the correct dynamical model describing signals cannot be specified a priori or when the exact underlying noise spectrum is not known. The POP approach, furthermore, is not optimized to guard against the biases (i.e., spectral leakage) known to exist in the spectrum estimation of finite time series (Thomson, 1982), and will provide unoptimal signal vs noise decompositions for relatively short and noisy time series.

Extended EOFs

Extended EOFs (Weare and Jasstrom, 1982; Graham *et al.*, 1987; Preisendorfer, 1988) identify the dominant spatiotemporal structure of lagged sequences of covariance estimates. Such a decomposition can thus capture time-evolving patterns in the data, since phase information is retained in the decomposition. The approach is useful to recover oscillatory patterns that are known to exist in the data, but cannot be used to detect spatiotemporal signals themselves without further generalization (see "Multichannel SSA" below).

Rotated EOFs + Spectral Analysis

Through selecting alternative rotated combinations of the eigenvectors obtained through PCA (e.g., "varimax rotation") (e.g., Richman, 1986; Houghton and Tourre, 1992), one can often obtain spatial patterns which may bear a closer relationship to particular, physically based modes of variability (e.g, the dynamical pattern of ENSO). To the extent that such rotation may allow for a more natural separation of the data into physically meaningful patterns, the PCA + SA procedure used above can be combined with a rotation procedure to provide a more faithful separation of the true signals in the data, and a more faithful signal detection procedure. Such a rotation is, however, subjective, requiring some a priori assumptions regarding the spatial patterns that are of physical significance (e.g., in the case of varimax rotation, that spatial structures of signals should be regionally localized). Objective selection rules for significance in PCA are also lost upon rotation. In some sense, rotation of EOFs prior to spectral analysis is an imperfect solution to the more fundamental problem that PCA provides only a time-domain optimal decomposition of the data variance, unable to appropriately recognize frequency-domain organization.

Complex Harmonic PCA

A complex generalization of PCA known as complex harmonic PCA or CH-PCA (see, e.g., Wallace and Dickinson, 1972; Barnett, 1983; Trenberth and Shin, 1984; Barnett, 1991; Preisendorfer, 1988; see also "principal components in the frequency domain," Brillinger, 1981, Chapter 9) provides a better description of oscillatory features in a multivariate data set than does conventional PCA. The CH-PCA procedure makes use of PCA on a matrix analogous to that defined by (2.7) but containing instead appropriately estimated complex spectral estimates $y^{(m)}$ of the data series

$x^{(m)}$ at all resolvable frequencies,

$$Y = \begin{bmatrix} y_{f_0}^{(1)} & y_{f_0}^{(2)} & \cdots & y_{f_0}^{(M)} \\ y_{f_1}^{(1)} & y_{f_1}^{(2)} & 2/ & y_{f_1}^{(M)} \\ \vdots & & & \\ y_{f_{N-1}}^{(1)} & y_{f_{N-1}}^{(2)} & \cdots & y_{f_{N-1}}^{(M)} \end{bmatrix}. \qquad (10)$$

A PCA is then performed in the transformed frequency domain,

$$Y = \sum_{n=1}^{N} \lambda_n \mathbf{u_n}^\dagger \mathbf{v_n}, \qquad (11)$$

where n in this context runs over the N distinct frequencies of the discrete Fourier transform of a data set of length N samples. The empirical orthogonal function (EOF) $\mathbf{v}_m$ describes the complex spatial pattern in amplitude and phase of the nth eigenvector, and the N-vector $\mathbf{u}_m$ now describes the combinations of pure harmonic components of variability that describe the relatively smooth time-evolution of the nth eigenvector. The eigenvalue (the square of the singular value) λ_m, as before, describes the associated fraction of data variance. Because phase information is maintained in this procedure, standing and traveling oscillatory signals in the data set are described more faithfully (see Preisendorfer, 1988, Chapter 12). The primary limitation of CH-PCA is that spectral estimates for the neighboring frequency estimates treated as independent random variables are in fact correlated, introducing a statistical dependence that is difficult to assess in interpreting the results of the eigendecomposition. Furthermore, modulated or irregular oscillations are not appropriately modeled.

Bandpass-Filtered PCA

In what can be viewed as an alternative to CH-PCA, a technique for identifying narrowband but anharmonic oscillatory features in the data is to prefilter with a bandpass over the frequency interval of interest (e.g., Trenberth and Shin, 1984), seeking to determine if there is a single dominant mode of variability within that restricted frequency band. While this approach does allow the detection of irregular narrowband oscillations, some of the more fundamental problems noted earlier for PCA + SA are not circumvented. In particular, because phase information is lost in the PCA, only standing oscillations can be described by any particular eigenmode. Furthermore, though bandpassing alleviates the most serious problems in PCA of successive operations of frequency-domain-optimized filtering (bandpassing) followed by time-optimal filtering (time-domain PCA of the data), the conventional filtering procedures invoked (e.g., the Hilbert transform—see Preisendorfer, 1988) provide less than optimal spectral resolution/variance trade-off properties (e.g., Thomson, 1982).

Multichannel SSA

The approach of multichannel SSA or M-SSA (Keppenne and Ghil, 1993; Allen and Robertson, 1996; Moron *et al.*, 1997), as in the method of extended EOFs described earlier, employs a multivariate correlation-space eigenvector decomposition to describe evolving spatially correlated structures in a multivariate data set. Indeed, the terminology of "extended EOFs" and "M-SSA" is sometimes used interchangeably (Allen and Robertson, 1996). We will draw a distinction, however, using "M-SSA" to describe the more general procedure of estimating the statistical significance (relative to a specified noise null hypothesis) of spatiotemporal oscillations detected in the lagged estimates of the data covariance matrix. Beyond detecting significant irregular spatiotemporal oscillations in a multivariate data set, M-SSA provides a direct link to the theoretical framework of nonlinear dynamical systems (see Vautard and Ghil, 1989; Ghil and Yiou, 1996). The approach provides an optimal decomposition in the correlation domain and not in the frequency domain (note the explicit comparisons of Thomson, 1982, of correlation-domain and frequency-domain estimators). SSA (and its multivariate counterpart M-SSA) can usefully analyze only those quasi-oscillatory structures with periods in the range $[L/5, L]$, where $L = N/3$ (see Vautard *et al.*, 1992), where L is the embedding dimension for the lagged-covariance estimation (equivalently, the width, in time units, of an equivalent moving window through the time series). Consequently, there are rather severe restrictions on the range of frequency bands over which temporal structure can be reconstructed simultaneously. For instance, to recover interdecadal patterns (approximate period 20 years), one may want to choose $L = 30$ years with a 100-year record. This window width will not allow the reliable decomposition of oscillatory signals (e.g., ENSO) with dominant time scales less than 6 years in this case. More importantly, in the multivariate context, the M-SSA approach runs up against severe dimensional limitations for large data sets. The introduction of multiple channels in the covariance estimation requires the statistical decomposition of a matrix in the time, spatial index, and lag domains. For a fixed-duration data series of length N and M channels (e.g., gridpoints), this requires the SVD of an $N \otimes ML$ matrix, which quickly becomes ill-posed (i.e., a unique eigendecomposition of the variance is not possible) as the number of spatial channels M becomes large. To avoid this problem, the spatial data set must first be further decomposed into a lower-dimensional representation (e.g., by conventional PCA) before the M-SSA algorithm is applied (Vautard *et al.*, 1992; Moron *et al.*, 1997). This latter step then tends to reintroduce some of the limitations of classical PCA noted above which we seek to avoid. In this sense, the usefulness of M-SSA becomes limited for spatially extensive data sets.

3. MTM-SVD Multivariate Frequency-Domain Climate Signal Detection and Reconstruction

The multitaper frequency-domain singular-value decomposition or "MTM-SVD" approach (Mann and Park, 1994; Mann *et al.*, 1995a, b; Mann and Park, 1996a, b) exploits the optimality of multitaper spectral analysis for analyzing narrowband signals superposed on a smoothly varying spectral noise background (see Section 2.1). The MTM-SVD approach seeks to isolate statistically significant narrowband oscillations (which may be modulated or "irregular" in nature) that are correlated among a sufficiently large number of normalized independent series or "channels" (e.g., multiple gridpoints) as to comprise a significant fraction of the total data variance. The approach invokes a null hypothesis of a smoothly varying colored-noise background, rejecting the null hypothesis when a large share of the multivariate data variance within a specified narrow frequency band can be attributed to a particular mode (i.e., modulated spatiotemporal oscillation) of variability. The approach can be appropriately modified with an "evolutive" generalization to describe broader-band and frequency-modulated processes (see Mann *et al.*, 1995b; Mann and Park, 1996b). Wavelet-based generalizations of the procedure more appropriate for the description of episodic variability have also been developed (Lilly and Park, 1995; Park and Mann, 1998).

As the MTM-SVD approach is complex-valued in nature, it naturally describes spatially correlated oscillatory signals with arbitrary spatial relationships in both amplitude and phase. In this manner, the approach can distinguish standing and traveling oscillatory patterns in a spatiotemporal data set. The multitaper decomposition also allows for a relaxation of the typically strict stationarity assumptions invoked in most spatiotemporal decompositions. The optimal frequency-domain properties of multitaper spectral analysis enable the procedure to provide superior signal detection and signal/noise separation under the assumption of narrowband signals and the null hypothesis of a spatially correlated colored-noise background with a smoothly varying spectrum. Moreover, because the methodology allows for the detection of either periodic or aperiodic irregular oscillatory patterns, it does not invoke restrictive assumptions regarding the governing dynamics. The characteristics of amplitude-, phase-, and frequency-modulated spatiotemporal oscillations assumed in the associated statistical model of "signal," for example, accommodate the description of stochastically excited linear climate oscillations and of self-sustained nonlinear oscillations equally well. Thus, the MTM-SVD technique provides a philosophical appeal over conventional multivariate techniques in an exploratory data analysis setting.

In this section, we describe details of the MTM-SVD method, including the techniques for signal detection, signal reconstruction, and confidence level and significance estimation. We demonstrate that the method provides the correct null inferences when applied to a class of spatially correlated colored- (red-) noise processes. Finally, we demonstrate highly successful spatiotemporal signal detection and reconstruction when the method is applied to the synthetic signal + noise example described in Section 2.1.

3.1. Signal Detection

The MTM-SVD signal-detection method makes use of the mutual information available from each of the K spectral estimates available at each frequency f in a multivariate data set of "spatial dimension" M. Rather than averaging the estimates of the distinct K eigentapers as in (2.5) and (2.6), the MTM-SVD approach retains the independent statistical information provided by each of the K eigenspectra, and seeks to find the optimal linear combinations of eigentapers that maximize the multivariate variance explained by a particular amplitude/phase modulation of a given carrier frequency component. The availability of multiple independent spectral estimates for each time series at a given frequency f is the fundamental requirement for the orthogonal decomposition employed in the MTM-SVD approach, and in almost all cases, the minimum value of the time-frequency bandwidth parameter $NW = 2$ is used, which admits ($K = 3$) such multiple degrees of freedom. This choice ensures minimal loss of frequency resolution. The reader is referred back to the discussion of Section 2.2 and to Figure 6. The decomposition describes a carrier oscillation modulated by a complex envelope function with K degrees of freedom, allowing for the description of modulated, irregular oscillations while providing the optimal spectral resolution/variance trade-off of multitaper spectral analysis.

We "standardize" each of the series to be analyzed by removing the mean over the N samples to yield an "anomaly" series $\{x'_n\}^{(m)}$ and normalize the resulting series by its standard deviation $\sigma^{(m)}$, where $n = 1, \ldots, N$ and N is number of samples (e.g., $N = 1200$ for 100 years of monthly data). This normalization favors the detection of spatially coherent processes. To represent the data in the frequency domain, we calculate the multitapered Fourier transforms for each normalized time series $x_n^{(m)} = x_n'^{(m)}/\sigma^{(m)}$,

$$Y_l^{(m)}(f) = \sum_{n=1}^{N} w_n^{(l)} x_n^{(m)} e^{i2\pi f n \Delta t} \quad (12)$$

for a given choice of K and the time-frequency bandwidth product $NW = p$, as in the univariate multitaper procedure (see (2.4)). Because secular variations are separated from higher-frequency variability with minimum spectral leakage, nonstationarity of the first order (e.g., a linear trend in the data) can be described with little bias on the rest of the spectrum, without any detrending or "prewhitening" of the data series. Thus, the decomposition avoids the strict stationarity requirements of most statistical time-series decompositions.

The $M \times K$ matrix,

$$\mathbf{A}(f) = \begin{bmatrix} w_1 Y_1^{(1)} & w_1 Y_2^{(1)} & \cdots & w_1 Y_K^{(1)} \\ w_2 Y_1^{(2)} & w_2 Y_2^{(2)} & \cdots & w_2 Y_K^{(2)} \\ \vdots & & & \\ w_M Y_1^{(M)} & w_M Y_2^{(M)} & \cdots & w_M Y_K^{(M)} \end{bmatrix} \quad (13)$$

is formed with each row calculated from a different gridpoint time series and each column using a different Slepian taper. The w_m represent spatially variable weights to adjust for relative areas of gridpoints, etc.

To isolate spatially coherent narrowband processes, a complex singular-value decomposition (e.g., Marple, 1987) is performed of the above matrix,

$$\mathbf{A}(f) = \sum_{k=1}^{K} \lambda_k(f) \mathbf{u_k}(\mathbf{f})^\dagger \mathbf{v_k} \quad (14)$$

into K orthonormal M-vectors $\mathbf{u}_k$, representing complex spatial empirical orthogonal functions (spatial EOFs), and K orthonormal K-vectors $\mathbf{v}_k$ (termed "spectral EOFs" by Mann and Park, 1994) which we will term here "principal modulations" in analogy with "principal components" of a time-domain eigendecomposition. Because the SVD is a multilinear decomposition, this approach posits a linear spatial relationship among all time series (e.g., spatial gridpoints) in any given signal. Any regional responses which are nonlinearly related to the large-scale signal may be imperfectly described by the estimated signal spatial pattern. The "principal modulations" describe the linear combination of projections of the K eigentapers which impose the amplitude and phase modulation of the oscillatory behavior associated with the kth mode. The key distinction between CH-PCA and MTM-SVD is that the MTM-SVD technique performs a *local* frequency-domain decomposition of K statistically independent spectral estimates as defined by (13), whereas CH-PCA performs a *global* frequency-domain decomposition over all spectral estimates (compare (10)). This distinction is the primary reason that the MTM-SVD technique can be used to isolate irregular oscillations superposed on an *arbitrary* smoothly varying colored-noise background.

The singular value $\lambda_k(f)$ scales the amplitude of the kth mode in this local eigendecomposition, where the K singular values are ordered $\lambda_1(f) \geq \lambda_2(f) \geq \cdots \geq \lambda_K(f) \geq 0$. The associated "eigenvalues" are the $\lambda_k^2(f)$. The normalized principal eigenvalue, $\lambda_1^2(f)/\sum_{j=2}^{K} \lambda_j^2(f)$, provides a signal-detection parameter that is *local* in the frequency domain. Under the assumption that no more than one signal is present within the narrow bandwidth of spectral estimation, the normalized principal eigenvalue should stand out distinctly above what would be expected from an appropriate noise model. In the relatively unlikely event that there exist two similarly strong signals within a single bandwidth, the usefulness of this detection parameter will be diminished. We refer to the normalized principal eigenvalue as a function of frequency as the "local fractional variance spectrum" or LFV spectrum. The LFV spectrum varies between $1/K$ and unity in magnitude, and has a variable frequency bandwidth $\Delta f_{\text{MTM-SVD}}$ between $\pm f_R$ and $\pm p f_R$, as it can be no more narrow than the Rayleigh resolution f_R and no greater than the bandwidth corresponding to a uniform average of the K eigenspectra (i.e., Δf_{BF} defined in Section 2.1). Correspondingly, only variability with period shorter than $N\Delta t/p$ (e.g., 50 years for 100 years of data and $NW = 2$) can be confidently distinguished from a secular variation. This multivariate spectrum provides a powerful frequency-domain signal-detection parameter, indicating the maximum fraction of narrowband spatiotemporal variance that can be explained by a particular modulated oscillation as a function of frequency. Typically, only this principal eigenvalue spectrum is used as a signal-detection parameter. An iterative procedure may be advised if there is reason to believe that multiple signals may be present in a particular narrow frequency band. For example, Mann and Park (1994; 1996b) use this latter procedure to identify two significant secular variations in instrumental climate data of the past century (see Section 4.1 for more details regarding this iterative procedure). As discussed below, the frequency-independent nature of the distribution of LFV for a wide range of colored-noise processes provides for fairly unrestrictive null hypothesis testing and the use of powerful nonparametric significance estimation procedures.

3.2. Signal Reconstruction

The spatial pattern of a signal associated with a significant peak in the LFV spectrum at frequency $f = f_0$ is described by the complex-valued M-vector, $\mathbf{u}_1$, which indicates the relative amplitude and phase of the signal at particular locations (e.g., gridpoints) of the multivariate data set. Using the envelope estimate $\tilde{A}_1(f_0, t)$ (see (17)), the reconstructed spatiotemporal signal $\tilde{y}$ is described by

$$\tilde{x}_n^{(m)} = \gamma(f_0) \Re \{\sigma^{(m)} u_1^{(m)} \tilde{A}_1(n\Delta t) e^{-i2\pi f_0 n \Delta t}\}, \quad (15)$$

where $u_1^{(m)}$ is the mth component of the spatial EOF $\mathbf{u}_1(f_0)$. The factor $\gamma(f_0) = 2$ for $f_0 \geq pf_R$, owing to contributions from spectral information near f_0 and $-f_0$. At $f_0 = 0$, $\gamma(f) = 1$. For $0 < f_0 \leq pf_R$, the value of γ is more problematic, as the sampling widths of the Slepian tapers in the frequency domain for f_0 and $-f_0$ overlap partially. In practice, it is simplest to treat such long-period variability as quasi-secular and use the $f_0 = 0$ passband for its reconstruction.

The canonical spatial pattern of the signal can be represented by the complex field

$$\hat{x}^{(m)} = \gamma(f_0) \sigma^{(m)} u_1^{(m)} A_{\text{rms}}(f_0), \quad (16)$$

where the pattern is scaled by the root-mean-square amplitude $A_{\text{rms}}(f_0)$ of the modulating envelope $\tilde{A}_1(f_0, t)$ (because of amplitude modulation, the amplitude of the pattern is variable from cycle to cycle). This reconstructed pattern describes the evolving spatial pattern over a cycle, and can thus be represented by a complex-valued pattern (see, e.g., Mann and Park, 1994), with the magnitude of the vector indicating relative amplitude and the angle indicating relative phase (i.e., relative timing of peak/minimum anomaly at a particular location for a particular variable). This information is often more physically portrayed in terms of a sequence of real-valued anomaly patterns (positive or negative values of the anomaly field) corresponding to the projection of the complex spatial vector onto various phases of a cycle (e.g., Mann and Park, 1996b).

The complex-valued K-vector $\mathbf{v}_1$ can be inverted to obtain the slowly varying envelope of the signal, similar to the complex demodulate. Park (1992) and Park and Maasch (1993) show how the slowly varying envelope

$A(t)$ of an oscillatory signal $x(t) = \Re\{A(t)e^{-i2\pi f_0 t}\}$ centered at a "carrier" frequency f_0 can be estimated from a set of eigenspectra $Y_l(f_0)$, $l = 1,\ldots,K$. The time-domain signal $x(t)$ and the envelope $\Re A(t)$ are formally identical for modes referenced to $f_0 = 0$, that is, the secular modes of variability. In the multivariate case, the time-domain signal is reconstructed from the components of its corresponding spectral EOF $\mathbf{v}_1^*(f_0)$. This reconstruction is not unique and requires additional constraints. The simplest reconstruction is an MTM version of the complex demodulate, a linear combination of the Slepian tapers $\{w_n^{(l)}\}_{n=1}^N$,

$$\tilde{A}_1(n\Delta t) = \sum_{l=1}^{K} \xi_l^{-1} \lambda_l(f_0) (v_1^{(l)})^* w_n^{(l)}, \tag{17}$$

where $v_1^{(l)}$ is the lth component of the vector $\mathbf{v}_1(f_0)$. The ξ_l are the bandwidth retention factors of the Slepian tapers (see Park and Maasch, 1993). This reconstruction tends to minimize the size of the envelope and thus favors $\tilde{A}_k \to 0$ at the ends of the time series. Such an inverse clearly is not appropriate for signals associated with nonstationarity in the mean (i.e., a secular trend). A second possible inversion minimizes the numerical first derivative of $\tilde{A}_1(n\Delta t)$ (Park, 1992), which favors envelopes that approach zero slope at the ends of the time series. Such an inversion does not discriminate against a zero-frequency trend in the data, for example, but is poorly suited for other describing features which change rapidly near the beginning or end of the data series. A third possible constraint minimizes the roughness of the envelope using the second derivative of $\tilde{A}_1(n\Delta t)$, which constrains neither the mean nor the slope near the ends of the data. Mann and Park (1996) favor a more general data-adaptive means of time-domain signal reconstruction in which the mean-square multivariate misfit with the raw data is minimized over all possible linear combinations of these three constraints. This approach removes the subjective reliance on some particular a priori boundary condition assumption and has been shown to provide optimal skill in a forecasting context (Rajagopalan et al., 1998).

For the evolutive procedure, the temporal reconstructions are performed separately in a sequence of staggered windows or "moving window" of specified width through the entire data series. The width of the window is typically chosen so that it includes multiple periods of the oscillatory signal of interest but is short enough to capture the evolution of frequency and amplitude features over the duration of the record. As an example, to study interannual (say, 3- to 5-year) oscillatory behavior associated with ENSO based on roughly century-duration records, we typically invoke a 40-year window in the evolutive procedure. The temporal reconstruction is in this case determined by a multivariate projection filtering (Thomson 1995; Mann and Park 1996a) using the reconstructions from overlapping intervals of adjacent windows. The spatial reconstruction for an evolutively determined signal (representing the "average" relative spatial phase and amplitude pattern of the signal) is determined by averaging the spatial patterns over multiple windows from the beginning to the end of the full data interval. In this case, the relative spatial patterns (in both amplitude and phase) vary somewhat over time, consistent with the possible nonstationarity of certain climate processes (see, for example, the discussion of secular changes in the characteristic effect of ENSO on precipitation patterns in certain regions by Ropelewski and Halpert, 1987). For such cases, either the spatial pattern corresponding to a particular window of time when the signal is strongest, or the average pattern over all windows, may be of most interest.

3.3. Testing the Null Hypothesis: Significance Estimation

The statistical significance of potential signals in the LFV spectrum requires an accurate estimate of the expectations from chance coincidence, given an appropriate null hypothesis. Following the earlier discussion of Section 2.1, the least restrictive null hypothesis that might be adopted is that the observed behavior arises from the statistical fluctuations of a spatiotemporal noise process with an arbitrarily colored noise background and a spatial correlation structure estimated empirically from the data set itself. The significance of putative narrowband signals detected in the LFV spectrum is estimated by diagnosing the likelihood that a given value of the LFV would arise from random fluctuations of such a process. Spatial correlation in the climate noise background which is largely local, but is, to a lesser extent, at the larger planetary wave scale also, strongly limits the true number of spatial degrees of freedom in the sampling of any climatic field. If such spatial correlation of the noise is not properly accounted for, incorrect significance-level estimation and spurious signal detection (i.e., peaks in the LFV spectrum) are likely. We guard against contaminations from long-range spatial correlations in the climate background noise by a bootstrap resampling of the multivariate data in time. This resampling destroys temporal, but not spatial, structure in the data set. Thus, MTM-SVD analysis of many independent time-resamplings of a multivariate data set can be used to calibrate the LFV confidence levels.

The LFV spectrum measures, within a narrow frequency band, the amplitude of the largest spatially correlated oscillatory "signal" relative to oscillations with other spatial patterns and temporal modulations. To be used as a signal detection parameter, the statistical significance of local peaks in the LFV spectrum is established by estimating the corresponding null distribution of the LFV parameter for spatially correlated colored noise in the absence of signal. The fundamental advantage of the LFV spectrum as a means of signal detection in this regard, is the universality of its underlying distribution for a very general class of noise processes. Under the assumption that the noise components of the time series that comprise the multivariate data set exhibit a smoothly varying colored-noise spectrum (e.g., as defined in Section 2.1), the null distribution is in fact independent of frequency and indistinguishable from that of white-noise series with the same underlying spatial correlation structure. This frequency-independence of the distribution results from the fact that the K eigenspectra estimated from a noise process will exhibit a Gaussian distribution at any frequency f as long as the noise spectrum appears flat over the narrow bandwidth $f \pm pf_R$ within which the eigendecomposition is actually performed. This local variance decomposition is resistant to influence from neighboring frequency bands owing to the spectral leakage resistance properties of MTM discussed in Section 2.2.

We demonstrate this frequency-independence of the null distribution of the LFV spectrum using the specific example of spatially correlated red noise, showing the virtual independence of the LFV spectrum of the level of temporal autocorrelation in the data set. There is nothing preferential about the AR(1) colored-noise model, however, and in fact the LFV exhibits a universal null distribution for any smoothly varying colored-noise process with a given number of spatial degrees of freedom. These features allow us to invoke a quite weak null hypothesis in signal detection that accommodates not only spatially correlated red noise, but in fact a wide class of spatially correlated colored-noise processes (e.g., the noise model of Wigley and Raper, (1990) discussed earlier). We exploit the frequency-independence of the null distribution by making use of a resampling technique for estimating this distribution and associated confidence intervals for significance. We employ a bootstrap method (Efron, 1990) in which the spatially distributed data set is resampled in such a way that the spatial patterns of the actual data are unaltered but their temporal sequence is randomly permuted in time. This permutation whitens the data set, destroying any temporal structure, but keeping all spatial structure intact so that the spatial degrees of freedom in the actual (e.g., monthly) data field are always faithfully represented. Because the distribution of the signal-detection parameter—the LFV spectrum—is independent of frequency under the assumptions described above, the distribution of the white bootstrap resamples is representative of that for any frequency. Within one bandwidth of zero frequency (the "secular band"), however, fewer spectral degrees of freedom are available because the Fourier transform becomes real-valued at $f = 0$. Within this secular band, the confidence levels must be estimated separately.

The null hypothesis described above can also be tested by an alternative "parametric" approach. If the number of spatial degrees of freedom M in the noise background can be reasonably estimated, an alternative parametric procedure for estimating the null distribution is provided by Monte Carlo simulation with M realizations of a Gaussian distributed process (see Mann and Park, 1994). Generally, the nonparametric bootstrap procedure is preferable, avoiding a priori assumptions of the spatial correlation structure of the noise. The frequency-independence of the distribution provides improved statistics on the quantile distribution, allowing the averaging of results over many independent frequencies of the discrete Fourier transform. Typically, 1000 independent bootstrap realizations are performed, providing good enough statistics for reliable estimation of the 99% threshold for significance. When dealing with monthly or seasonal data in which there may be seasonal inhomogeneity in the variance as well as in the mean (for example, different seasons may have different levels of temporal autocorrelation—see, e.g., Briffa and Jones, 1993), it is often advisable to perform the bootstrap procedure separately for each month of the year, averaging the results for all months or the appropriate season analyzed (see Mann and Park, 1996b).

To test the proper distribution estimation of the bootstrap procedure, we generated replicates of the synthetic spatiotemporal red-noise process described in Section 2.1 with varying levels of temporal autocorrelation. The same white-noise innovation was used in each case, so that the stochastic element is identical for each case. The time series for the reference center gridpoint are shown for increasing values of ρ (Fig. 10a). The case $\rho = 0$ corresponds to a pure white-noise sequence. The lower-frequency variability becomes relatively enhanced as ρ increases, with the case $\rho = 0.9999 \approx 1$ nearly a nonstationary random walk (the reader is referred back to Fig. 2).

We applied the MTM-SVD methodology with $K = 3$ and $NW = 2$ to estimate the LFV spectrum for each spatiotemporal noise realization, estimating significance levels from the bootstrap procedure described above. The estimated null distribution, the reader might note, will be independent of the value of ρ owing to the whitening nature of the bootstrap. Of interest, then, is whether this independence holds up, at least under reasonable degrees of redness, for the observed distributions of LFV spectra for the red-noise processes themselves. The LFV spectra for the different cases are shown in Figure 10b. Since the case $\rho = 0$ corresponds to white noise, the null hypothesis of a smoothly varying

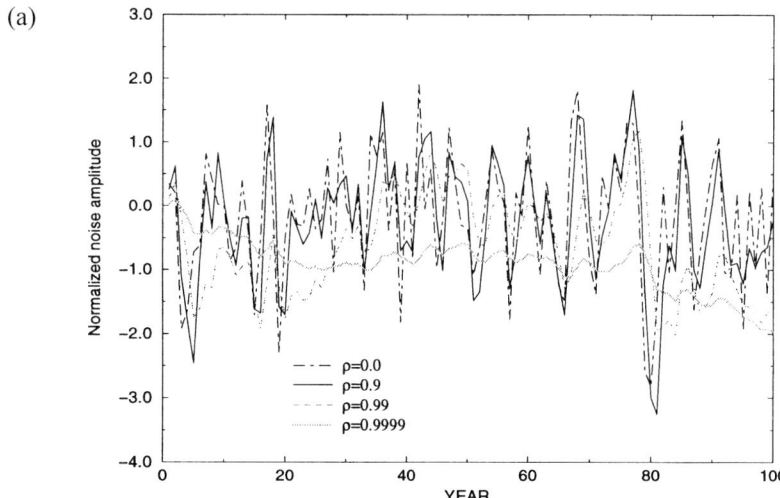

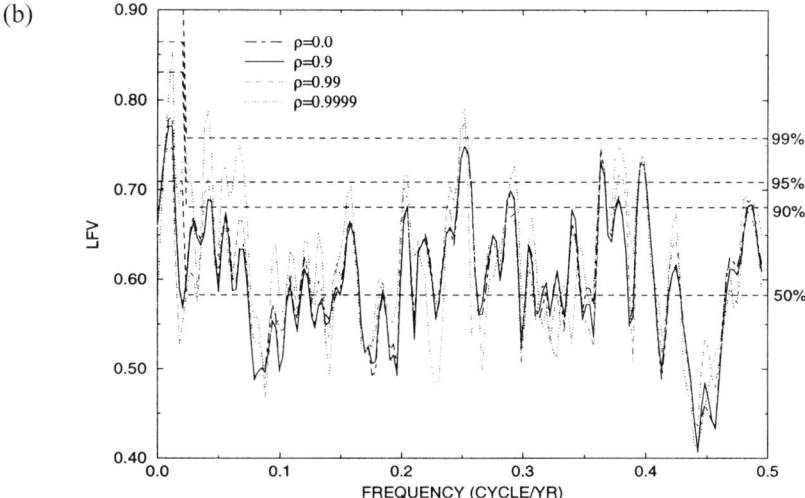

FIG. 10. (a) Center gridpoint annual mean reference series for spatiotemporal noise realization with varying levels of autocorrelation ρ. (b) LFV spectrum of the MTM-SVD analysis based on multivariate spatiotemporal noise realization with varying levels of autocorrelation ρ. [From Mann (1998).]

"locally white" noise background should trivially be satisfied. Indeed, the observed LFV spectrum breaches the 99% confidence level at a 0% rate, the 95% level at a 4.1% rate, and the 90% level at an 11.7% rate. This distribution is consistent with the expected rates of chance coincidence for rejecting the null hypothesis (1%, 5%, and 10%, respectively). Moreover, the distribution for moderate red noise (i.e., $\rho = 0.9$, redder in fact than observational climate data as discussed in Section 2.1) is virtually indistinguishable on the scale of the plotted LFV spectrum from that of the pure white-noise case (compare cases $\rho = 0$ and $\rho = 0.9$ in Fig. 10). Even for the quite strongly red case $\rho = 0.99$, the observed spectrum is quite close to that for the white-noise case. Only as the red-noise spectrum nears singularity (i.e., $\rho = 0.9999 \approx 1$) does the distribution of the LFV spectrum noticeably depart from that of the pure white-noise case. The most obvious discrepancies are observed at low frequencies where the parameter F defined in Section 2.1 far exceeds unity. A more thorough experiment employing ensembles of 1000 random trials (Table II) demonstrates that the correct rates of chance exceedance of given confidence levels are indeed obtained for all but the largest value of ρ. Thus, the null distribution of the LFV spectrum has been shown to conform to the pure white-noise distribution under precisely those conditions for which our a priori definition of a smoothly varying colored-noise spectrum is satisfied. While we have demonstrated the frequency-independence of the null distribution for the LFV spectrum and the validity of the significance estimation procedure for the case of a smoothly varying red-noise spectrum, we assert without demonstration that the requirement is much more generally just that of smoothly varying colored-noise background, for which the spectrum does not vary abruptly with frequency (i.e., a "locally white" noise background). This generality of the null hypothesis invoked in the signal detection procedure is a significant strength of the MTM-SVD methodology.

TABLE II RATES OF EXCEEDANCE OF A GIVEN CONFIDENCE LEVEL FOR SIGNIFICANCE AS A FUNCTION OF THE NOISE AUTOCORRELATION LEVEL ρ. [FROM MANN (1998).]

ρ	50%	90%	95%	99%
0.0	50%	10%	5%	1%
0.9	50%	10%	5%	1%
0.99	50%	12%	6%	2%
0.9999	50%	15%	10%	5%

3.4. Application to Synthetic Data Set

We now apply the MTM-SVD methodology to the full synthetic (spatiotemporal signal + noise) data set described in Section 2.1 using $K = 3$ and $NW = 2$. Figure 11a shows the LFV spectrum of the full 100 years of synthetic monthly data. Each of the a priori signals (secular trend corresponding to the zero-frequency peak, interdecadal peak centered near $f = 0.065$ cycle/yr, and multiple peaks within the $f = 0.5$- to $f = 0.33$-cycle/yr band of the frequency-modulated interannual signal) are significant well above the 99.5% level. There are no spurious peaks at the 99% level, consistent with chance expectations. With $NW = 2$ and $N = 1200$ months of data, there are between 25 and 50 independent values of the LFV spectrum within the range $f = 0$ to $f = 0.5$ cycle/yr, so that roughly speaking, no spurious peaks are expected at the 99% level, and only one or two at the 95% level.

The multiple, closely spaced set of highly significant peaks in the $f = 0.2$- to $f = 0.33$-cycle/yr (3- to 5-year-period) range that were detected in the LFV spectrum are suggestive of a more broadband signal. Thus, it is useful to use an evolutive version of the analysis to see if a more parsimonious description of the signal is evident. Figure 11b shows the evolutive LFV spectrum based on a 40-year moving window. Note that in this case, only oscillatory signals with period shorter than 20-year time scale can be resolved from a secular variation, so that the interdecadal signal and the secular trend appear as a single merged low-frequency streak in the evolutive spectrum. In the resolvable interannual band, however, a single dominant band of significant variance emerges, drifting from a dominant frequency $f = 0.35$ cycle/yr ($\approx$ 3-year period) early, to $f = 0.5$ cycle/yr ($\approx$ 2-year period) near the end of the series. Amplitude modulation is also somewhat evident in the evolutive LFV spectrum itself, although a reliable estimate of the amplitude modulation is only possible through signal reconstruction. It is thus clear that the description of a single frequency-modulated interannual signal provides a more parsimonious description of the group of peaks in the interannual band found in the spectrum of the full 100-year series.

Each of the three signals is independently reconstructed (Fig. 12). The secular and interdecadal oscillatory signals are reconstructed by the standard technique described in Section 3.2, while the interannual signal is reconstructed based on the evolutive reconstruction method, employing a 40-year moving window. Both the spatial and the temporal reconstructions are faithful to the exact counterparts (Fig. 3), resolving much of the complicated spatial variations in amplitude and phase of the true signals. Since the signals were immersed in spatially correlated red noise, some

(a)

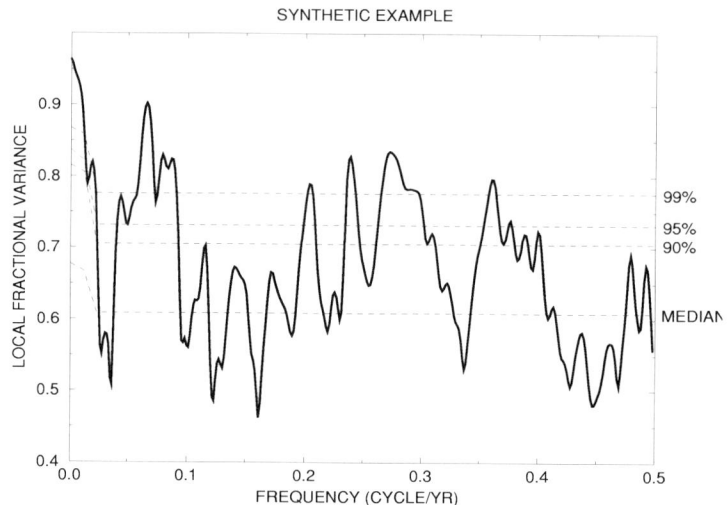

(b)

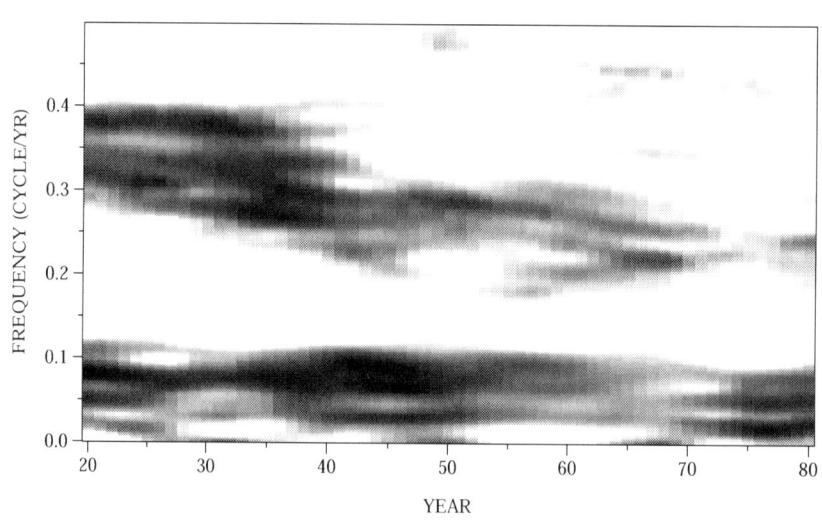

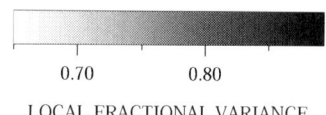

degree of noise contamination is unavoidable, and small spatially correlated errors in both amplitude and phase are evident in the reconstructed spatial patterns. It is worthy of note that the spatial correlation structure of the noise itself can lead to spurious small-scale coherent departures from the true signal. Nonetheless, the phase and amplitude errors are modest, and it is clear that reconstructed signal amplitudes are very small in those regions where no original signal was present (compare, e.g., the spatial patterns for the true and the reconstructed interdecadal and secular signals at the nodes of the true signals). More simplistic means of signal projection (e.g., the common bandpassing of a multivariate data set over a preferred range of frequencies) will lead to considerable spurious projection of the noise background. Such errors are largely avoided in the MTM-SVD signal reconstruction procedure, because only a particular modulated component of the narrowband variance is projected out in signal reconstruction. Consequently, regional errors in amplitude and phase are smaller.

These observations underscore a shortcoming of conventional PCA that is overcome in the MTM-SVD approach. As discussed previously in Section 2.2, the spatial patterns of PCA eigenmodes are often arbitrary, and it is thus difficult to distinguish true global signals from combinations of regional signals artificially combined through the PCA procedure. Procedures such as varimax rotation (Richman, 1986) may yield more physically sensible patterns, but only under certain restrictive assumptions regarding the spatial structure of signals; in fact, the true linear transformations of the eigenmodes required to yield physically distinct climate signals cannot a priori be known. In the case of MTM-SVD, a corresponding alternative linear transformation—the Fourier transform—is in fact specifiable a priori. The MTM-SVD procedure assumes that a signal has a very specific narrowband frequency-domain structure, and under that assumption assures quite high spatial signal-to-noise ratios in signal reconstruction. While there is considerable power in the synthetic data (signal-plus-noise) at all gridpoints at all frequencies, the regional as well as the large-scale spatial structure of the reconstructed signals were shown above to be quite faithful to their true counterparts. The ambiguities in distinguishing regional vs global spatial structures, inherent in conventional PCA, are thus relatively alleviated in the MTM-SVD approach. Similarly,

FIG. 11. Local fractional variance spectrum of SVD based on (a) full 100 years of monthly data and (b) evolutive analysis with a 40-year moving window. In the former case, 90%, 95%, and 99% confidence levels are shown with dashed line. For the latter case, the LFV spectrum amplitude is indicated in grayscale, filtered at the 90% level for significance. Darker contrast indicates greater amplitude and significance. [From Mann (1998).]

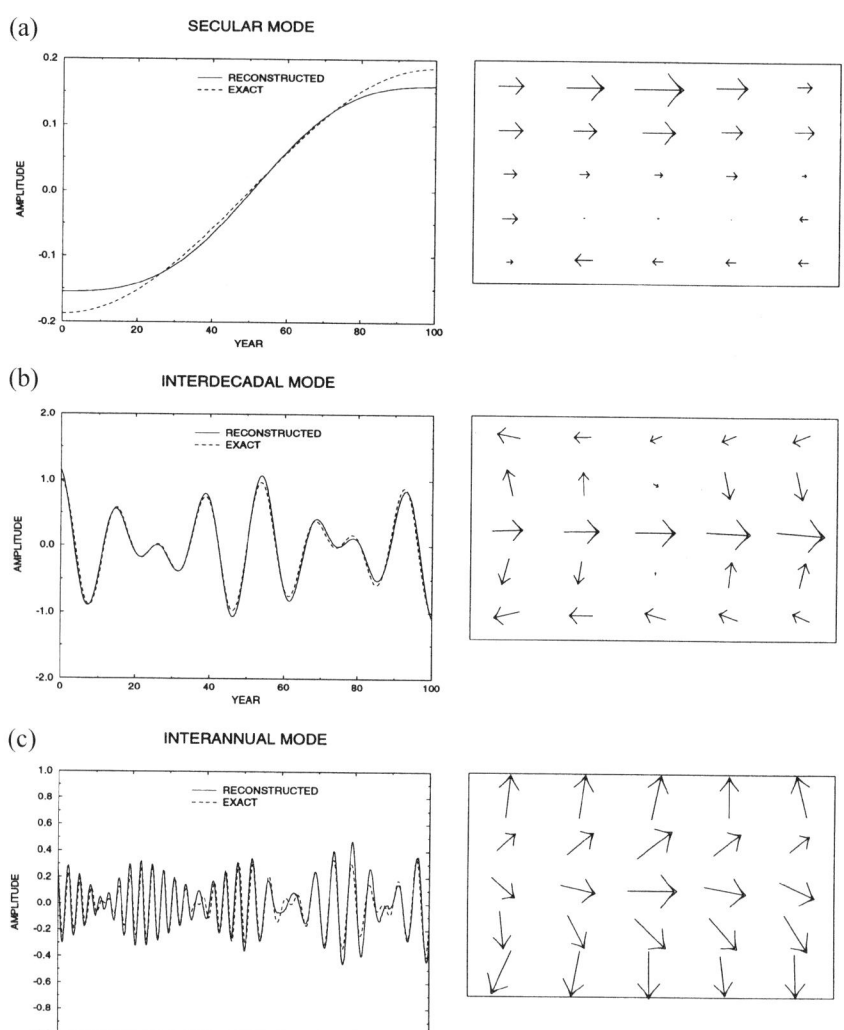

FIG. 12. Temporal (left) and spatial (right) patterns of reconstructed synthetic signals showing (a) secular mode, (b) interdecadal mode, and (c) interannual mode. Actual reference time-domain signal (see Fig. 2) is shown by dashed curve for comparison to reconstructed signal. [From Mann (1998).]

as discussed later, some of the deficiencies inherent in joint-field decompositions are avoided when performed in the context of MTM-SVD. In the event where more than one signal is present *within* the narrow bandwidth of spectral estimation, however, problems and ambiguities similar to those encountered in traditional PCA can arise. Such an example is presented and discussed in Section 4.1.

3.5. Effects of Sampling Inhomogeneities

Finally, we test the sensitivity of the MTM-SVD methodology to the sorts of potential sampling problems encountered in actual climate data sets. We examine the effects of temporal gaps in the sampling, as well as the effects of the sparse spatial subsampling of the global domain. We test the sensitivity to inhomogeneous time sampling by placing random gaps independently in each of the 25 gridpoints of the synthetic data set, introducing a proportion of missing data that varies linearly from 50% at the beginning of the $N = 1200$ months to 0% at the end. This trend in missing data simulates the gaps in instrumental climate data which are far more prevalent early in the instrumental climate record and virtually absent in the most recent record. In the true climate record, this trend in sampling is somewhat more complex (for example, World War II induced a sudden decrease in spatial sampling of climate observations—see, e.g., Bottomley *et al.*, 1990). The precise dates of the missing data are not correlated in space, again consistent with the typical missing data bias found in actual climate data. The gaps introduced as described above are somewhat more severe than those found in any of the gridded temperature data sets described earlier, ensuring a conservative test of the impact of missing data in actual climate records. The MTM-SVD analysis is repeated with this missing data, filling gaps with linear interpolation of neighboring values in time for each gridpoint (Fig. 13). The missing data has little discernible effect on the analysis, owing largely to the fact that the serial correlation in time due to both signal and noise decreases the amount of independent information in any one sample. Given that the synthetic data has similar temporal correlation structure to the observational climate data, we conclude that sizable rates of randomly missing data have little influence on the ability to discern significant frequency-domain information in the multivariate data. The impact of systematic biases in the collection of data, hidden in the processing stage of gridded monthly data typically used for climate studies, exhibits a greater potential to decrease the fidelity of these records, but considerable attempts have been made to document and ameliorate the impacts of such biases in gridded climate

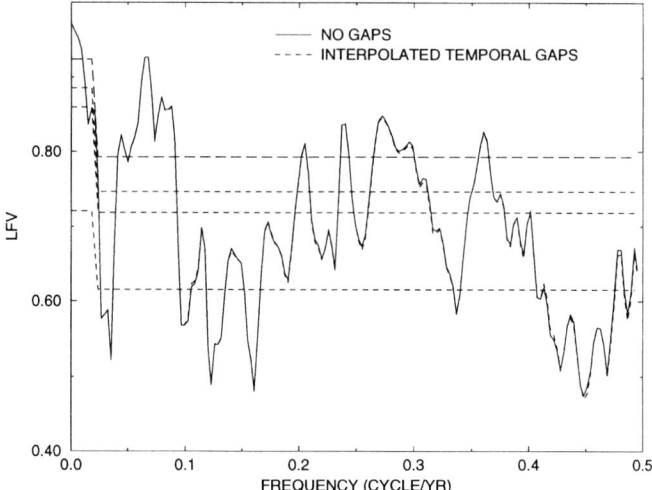

Fig. 13. Comparison of LFV spectrum for complete data set and for the case where temporal gaps are present (see text) which are linearly interpolated. [From Mann (1998).]

data (see the review by Parker *et al.*, 1995). Jones and Briffa (1992) favor the use of sea surface temperature (SST) rather than marine air temperature (MAT) measures of maritime temperature variations owing to historical changes in the diurnal timings of sampling in the latter case. SST measurements, on the other hand, suffer from systematic biases due to changes in bucket collection methods that can, if imperfectly, be estimated (Bottomley *et al.*, 1990).

Finally, we examine the bias of inhomogeneities in the spatial sampling of climate data by employing sparse subsamples of the full spatial domain (see Fig. 14) to the data set described above. The application of the MTM-SVD methodology to these sparser spatial networks reveals a surprising insensitivity to the precise subsampling of the spatial domain, though the relative prominence and detectability of signals depends on whether or not regions where the signal is strongest are included in the spatial network (Fig. 15). The "checkerboard" network grid of case I containing 13 of the 25 gridpoints exhibits an LFV spectrum which is virtually indistinguishable from that of the full grid. So too does the more regionally restricted network of 10 gridpoints of case II, with no differences in peak detection at the 95% or greater levels of inference. The sparser regular network of 9 gridpoints of case III begins to show signal-detection degradation, with the interannual peaks detected at a lesser (95–99%) degree of confidence than in the full data set and cases I and II.

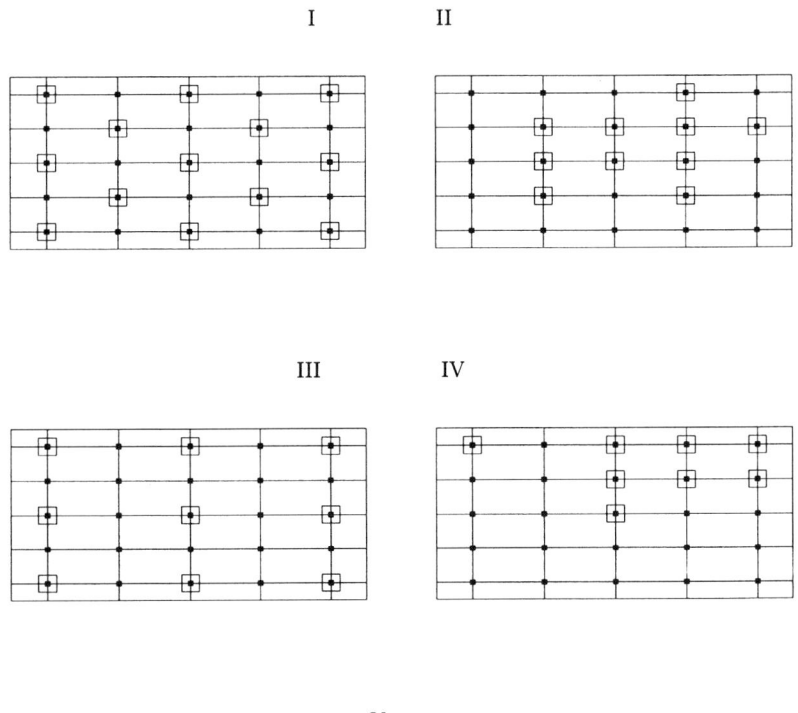

FIG. 14. Various spatial subsets of the domain used for testing sensitivity to spatial sampling. [From Mann (1998).]

Similar observations hold for the "Northern Hemisphere"-only sparse network of 8 gridpoints in case IV. In this case, the sparseness of "tropical" sampling, where the interdecadal signal is most prominent, leads to decreased detectability of the signal; it is just barely isolated at the 99% level of significance. Only with the very sparse network of 5 gridpoints in case V does the signal detection procedure suffer markedly. This network, for example, only samples the grid where the secular trend either vanishes or is relatively weak. Not surprisingly, the secular peak is not isolated as

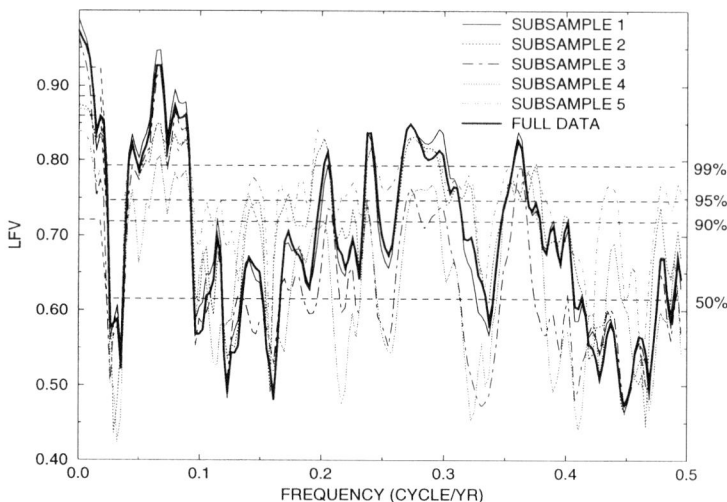

FIG. 15. Comparison of LFV spectrum for complete spatial network along with the various spatially subsampled networks described in Figure 14. The numerical vertical scale shown applies only to the full 25-gridpoint data set. The LFV spectra for the various spatial subsets are renormalized so that the quantiles of the null distribution and significance levels shown roughly apply to each of the subsets, irrespective or the varying spatial degrees of freedom present. [From Mann (1998).]

significant. Furthermore, the sampling network contains only a small number of spatial samples, and they are distributed over a small subregion where the relatively larger coherence scale of the signals relative to noise cannot be as readily exploited. Moreover, with so few spatial degrees of freedom, the temporal gaps introduced in the individual synthetic series become more problematic, as there is little mutual information in space to help guide the spatiotemporal decomposition. In this case, distinctions between signal and noise at, e.g., the 95% confidence level are less decisive. Notable spurious peaks (e.g., two between $f = 0.4$ and $f = 0.5$ cycles/yr) breach the 95% level of significance. Comparing with LFV spectra of the denser grids (I–IV) allows us to visualize how increasing degrees of freedom in sampling allow for more clearcut signal/noise separation, damping out these noise fluctuations.

While a precise comparison of the spatial gaps in the synthetic data to those in the observational data is not possible because the relevant spatial scales and densities of sampling are not directly comparable, these results suggest that the samplings available in the instrumental record are probably adequate to analyze signals representative of global or hemispheric domains. The sparse networks of long-term proxy climate indicators available for analysis (Mann *et al.*, 1995b) are more likely to suffer some of the biases evident, for example, in the worst-case scenario (V) discussed above.

4. Applications of MTM-SVD Approach to Observational and Model Climate Data

The search for oscillatory signals in the climate record exhibits a long and sometimes *checkered* history. Other than certain climate processes, such as the 3- to 7-year El Niño/Southern Oscillation (ENSO) for which the underlying dynamics are now relatively well understood (see, e.g., Cane *et al.*, 1986; Philander, 1990), the detection of oscillatory signals in the climate record has remained controversial (see, e.g., the review by Burroughs, 1992). Increasingly widespread and higher-quality climate data and the development of more sophisticated statistical analysis techniques have led to more confident exploratory signal detection in climate studies. Several recent studies, for example, have provided evidence for decadal- and longer-time-scale oscillatory climate signals in greater than century-long records of estimated global- or hemisphere-mean surface temperature (e.g., Folland *et al.*, 1984; Kuo *et al.*, 1990; Ghil and Vautard, 1991; Allen and Smith, 1994; Mann and Park, 1993; Schlesinger and Ramankutty, 1994; Mann and Lees, 1996). Without providing a spatial picture of variability, however, such studies shed little insight into the possible processes that may be responsible.

Simultaneous analysis of multiple indices of climatic variability, including vertically resolved oceanographic and atmospheric data (see, e.g., Wallace *et al.*, 1992; Fraedrich *et al.*, 1993; Xu, 1993) have the potential to offer the most insight into underlying dynamical processes, but long-duration (i.e., century-long) globally distributed records are not available. Geopotential height data, for example, are available for only a few decades. Long records of sea-level pressure are available but are largely confined to the Northern Hemisphere. Widespread records of precipitation exist, but they represent a more indirect proxy for underlying physical processes. Global temperature data provide widespread coverage for almost a century, and probably provide the greatest potential for the detection of interannual and decadal-scale spatiotemporal climate signals. Only proxy climate data, however, can provide a longer-term perspective on multidecadal- and century-scale climate variability. Qualitative studies of longer-term proxy climate data (e.g., Bradley and Jones, 1993) have been undertaken, but systematic multivariate analyses of these data are at a preliminary stage (see Bradley *et al.*, 1994; Diaz and Pulwarty, 1994; Mann *et al.*, 1995).

It is thus worthwhile to analyze all of the complementary data available, both instrumental and proxy, to isolate persuasive evidence for signals in the climate record. Few early studies analyzed records of sufficient duration and global extent to characterize modes of climatic variability at decadal and longer time scales. Furthermore, most applications of conventional spatiotemporal signal-detection approaches to climate data have suffered the limitations outlined in Section 2.2. Seeking to obtain a clearer picture of possible low-frequency signals in the climate record, we review in this section the application of the MTM-SVD multivariate signal-detection approach described in Section 3 to various instrumental and proxy climate data sets.

We first describe an analysis of globally distributed monthly land air and sea surface temperature data available during the past century (see Mann and Park, 1994). The spatiotemporal nature of the analysis allows us to judge the relative importance of spatially uniform variations which may be associated with changes in the global surface heat budget, and more regionally heterogeneous patterns which may be indicative of the relocation of heat by anomalous patterns of atmospheric circulation. Aside from identifying well-established quasi-biennial and interannual ENSO signals in the data, this analysis provides evidence for less well-established decadal and multidecadal signals. These signals include a 15- to 18-year time-scale oscillatory pattern exhibiting important tropical and extratropical features, and a secular "multidecadal" variation associated with a single cycle of warming and cooling global in extent but most pronounced in the North Atlantic.

To obtain a more direct picture of the possible dynamical processes governing such signals, we analyze spatiotemporal variability jointly between surface temperature and sea-level pressure (SLP) records that are available for nearly a century covering much of the Northern Hemisphere (see Mann and Park, 1996b). This analysis yields independent evidence for the signals discussed above and provides physical insight into possible underlying dynamical processes. The coupled oscillatory patterns of surface temperature and atmospheric circulation anomalies provide a more specific "fingerprint" of variability for comparison with signals found in climate model simulation studies. Seasonal and time-dependent features of the signals are more closely examined in this analysis. While sacrificing the global scope of the temperature-only analysis, the joint-field analysis provides more dynamical insight, and nearly complete spatial sampling of the Northern Hemisphere region.

Next, to address the difficulty in isolating multidecadal and longer-term oscillations in the short instrumental record, we analyze a globally distributed set of disparate proxy ("multiproxy") temperature records of several centuries duration (Mann et al., 1995b). This analysis provides evidence for persistent 15- to 30-year-period interdecadal, and 50- to 100-year century-scale climatic oscillations. While the widespread sampling available in the instrumental record is not available in the proxy data sampling, the resolvable features of the spatial patterns appear to be consistent with their instrumental counterparts. Most importantly, information regarding the long-term amplitude and frequency modulation and the persistence of oscillations over time is available from this analysis.

Finally, we focus the MTM-SVD method on the frequency band centered on $f = 1$ cycle/yr, i.e., the yearly cycle of temperature, to examine historical shifts in the timing of the seasonal cycle. To do this, we restore the yearly cycle of temperature at each gridpoint to the temperature anomaly data set described by Briffa and Jones (1992). We verify that the seasonal shift reported by Thomson (1995) is present in this data set and appears strongly concentrated in the continental interiors of the Northern Hemisphere continents. We show that the observed spatial pattern of seasonal shifts is in conflict with predictions of the effects of enhanced-greenhouse conditions in two well-known numerical climate models.

4.1. Global Temperature Data

Global surface temperature records provide a long and widespread sampling dating back to the nineteenth century. Such data sample most of the globe, albeit quite sparsely in certain regions. Mann and Park (1993) demonstrated that interannual and interdecadal temperature signals, while widely correlated, exhibit spatial variability that leads to considerable cancellation in a global average. Thus, analyses of hemisphere- or global-mean surface temperature alone (e.g., Folland et al., 1984; Kuo et al., 1990; Ghil and Vautard, 1991; Allen and Smith, 1994; Mann and Park, 1993; Schlesinger and Ramankutty, 1994; Mann and Lees, 1996) can yield at best incomplete information regarding low-frequency climate signals. An analysis of the spatial patterns of signals is essential to capture more fully the complex regional variations in amplitude, sign, or phase of low-frequency global temperature signals. Here we seek to isolate the full spatiotemporal structure of oscillatory modes of variability in global temperature, including those whose effect is largely to redistribute heat over the Earth's surface. Such signals would scarcely be evident in large-scale temperature averages.

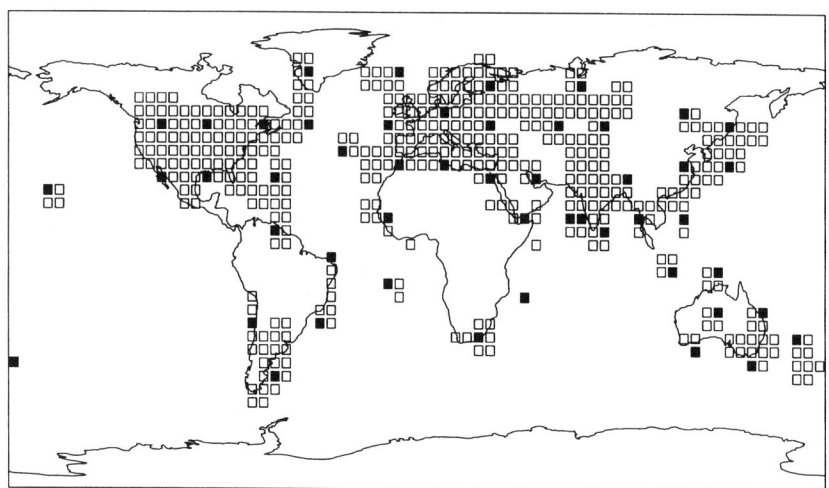

FIG. 16. Locations of 5° × 5° gridpoints used in analysis. The filled boxes indicate the sparse, more homogeneously distributed subnetwork of 50 gridpoints used to test sensitivity to spatial sampling.

The temperature data used in this analysis (Fig. 16) consist of land air and sea surface temperature anomalies distributed on a 5° × 5° global grid (see Jones and Briffa, 1992). To obtain nearly continuous monthly sampling from 1891 to 1990, we use a subset consisting of $M = 449$ gridpoints containing only small gaps (less than 6 months). We interpolated these gaps linearly, yielding time series of length $N = 1200$ months (i.e., 100 years). Such interpolation is defensible in light of the demonstrations regarding temporal inhomogeneity in Section 3.5. While notable spatial gaps are evident over certain regions (e.g., the southern oceans, large portions of the North Atlantic and Pacific, and Africa), the most seriously unsampled regions—the high latitudes—represent a small proportion of the global surface area. In light of the tests of spatial sampling sensitivity described in Section 3.5, the available gridpoint data should be sufficient for establishing global-scale signals. Nonetheless, as described later, we provide an additional internal consistency check by comparing results from a relatively homogeneous, sparse spatial subsampling of the data with those of the full gridpoint data set (see also Fig. 16).

An attempt has been made (Jones and Briffa, 1992) to remove the potential sources of systematic bias in this data set arising from urban warming, historical changes in data collection, and the weighting of data within gridpoints. To the extent that some residual biases are inevitable, we refer the reader to discussions by those who have looked into these issues most carefully (Jones and Briffa, 1992; Parker et al., 1995; Bottomley et al., 1990). Similar applications of the MTM-SVD methodology to spatially interpolated instrumental climate data sets of greater than century duration (e.g., Kaplan et al., 1997) have recently been undertaken (Tourre et al., 1998).

Here we apply the standard MTM-SVD analysis procedure described in Section 3 to the temperature data set, with the conventional choices $K = 3$ spectral degrees of freedom and bandwidth parameter $p = 2$ that were advocated previously. The gridpoint anomaly series are standardized (i.e., the long-term mean is removed and the residual is normalized by its standard deviation). The $M = 449$ gridpoints are uniformly weighted in the analysis. Temporal signal reconstruction is performed based on a priori specified boundary constraints of "minimum-slope" for secular variations and "minimum-norm" for oscillatory signals (see Section 3.2). Justification of these choices is provided by Mann and Park (1994), although more objective boundary constraints (see Section 3.2) are used in the subsequent analysis of joint SLP and temperature data (Section 4.2).

LFV Spectra

Figure 17 shows the LFV spectrum resulting from the MTM-SVD analysis of the 100-year data set over the broad frequency range $f = 0$ to $f = 2.0$ cycle/yr (i.e., periods from secular trend through half-year). Statistical significance levels shown were taken from the bootstrap resampling estimates of the null distribution. A separate parametric analysis (Mann and Park, 1994) suggests that this distribution is equivalent to that of Gaussian (locally) white noise with $\tilde{M} = 20$ spatial degrees of freedom. The most prominent peaks are the secular ($f \approx 0$) peak and those corresponding to the seasonal cycle and its first harmonic at $f = 1$ and $f = 2$ cycle/yr. The latter peaks are somewhat unexpected in *deseasonalized* temperature anomaly data for which the seasonal cycle has nominally been removed (Jones and Briffa, 1992). Due to the deseasonalization process, one statistical degree of freedom has been removed from any variability at the annual cycle or its harmonics. The decreased number of degrees of freedom raises the required levels of significance in the LFV spectrum from those shown, but does not alter the conclusion that the annual cycle and its first harmonic are significant in the anomaly data. This anomalous behavior in the annual cycle of global temperature anomaly data was first noted by Kuo *et al.* (1990), and appears to be associated with low-frequency changes in the timing of the seasons (Thomson, 1995; Mann and Park, 1996a). Otherwise, the LFV spectrum exhibits for the most part the

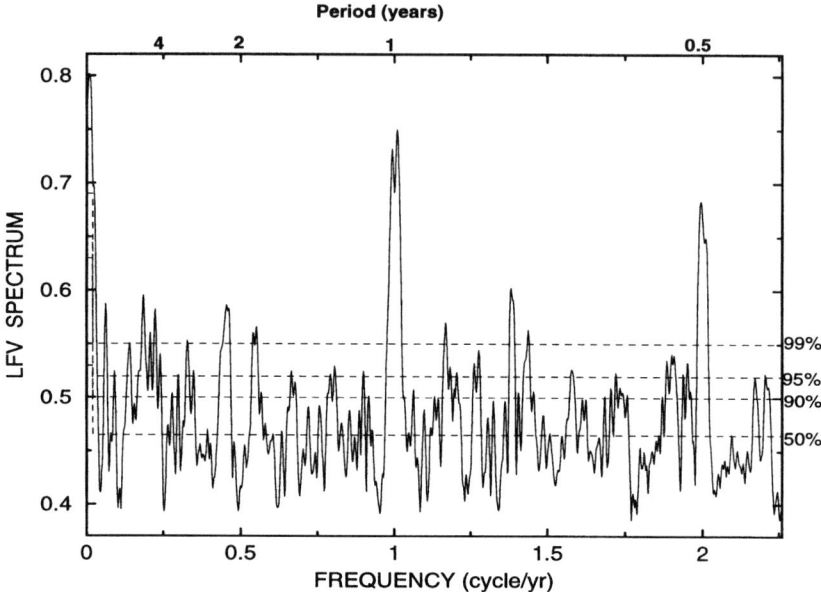

FIG. 17. LFV spectrum of the global temperature data shown in Figure 16 through frequencies slightly greater than $f = 2$ cycle/yr. Horizontal dashed lines denote 90%, 95%, and 99% confidence limits from bootstrap resampling.

frequency-independent spectrum expected for a colored-noise process (the reader is referred back to the discussion of Section 3.3) but with a somewhat greater number of 99%-significant peaks (11 aside from the 3 discussed above) than would be expected (1–2 following the discussion in Section 3.3) from chance coincidence alone. Of these, 8 are found in the interannual ($f < 0.5$ cycle/yr) band, corresponding mostly to frequencies (e.g., the ≈ 2.1-year quasi-biennial period and the 3- to 7-year-period ENSO band) associated with a priori established climate signals. Other apparent signals, however, are more disputable. Note that the background LFV spectrum (e.g., the depth of the noise floor) does not exhibit a corresponding increase in the interannual frequency range, underscoring the fact that the distribution of the LFV spectrum is similar at the low frequencies, as expected, but that there are simply a greater number of excursions at the highest percentiles. This observation is consistent with the detection of a small number of narrowband interannual signals superimposed on the noise background. We focus on these below.

Figure 18 shows the LFV spectrum in the interannual ($f > 0.5$ cycle/yr—i.e., periods greater than 2 years) range. Results for the full 100 years

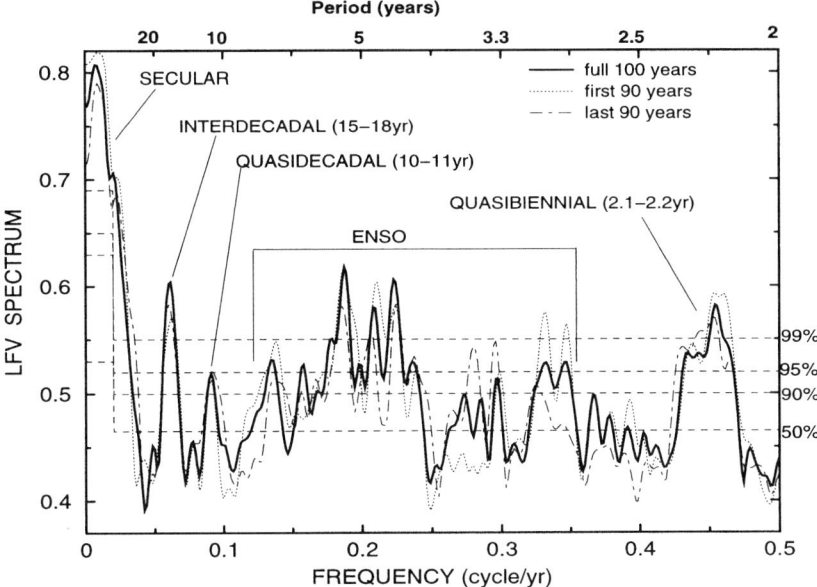

FIG. 18. LFV spectrum of the global temperature data set in the restricted "interannual" range $f = 0$ to $f = 0.5$ cycle/yr. The spectra for all 100 years (solid), first 90 years only (dotted), and last 90 years only (dot-dashed) are each shown for comparison. Horizontal dashed lines denote 90%, 95%, and 99% confidence limits from bootstrap resampling. Putative signals are noted by the indicated labels. [From Mann and Park (1994).]

of monthly data are shown along with those based on only the first and last 90 years of data. The latter calculations provide a test of the robustness of signals "found" in the LFV spectrum. Further truncation of the data series would decrease the spectral resolution of the LFV spectrum to the point where meaningful comparison is not possible. The irregular nature expected of the signals will lead to variations in the relative prominence of individual spectral peaks over different time intervals, but consistency among the three trials allows more confident signal-detection inferences. The LFV spectrum indicates peaks centered at 2.2 years, several peaks within the 4- to 6-year band, a peak centered at the 15- to 18-year period, and a secular peak where the significance breaches the 99% confidence level for nonrandomness in each of the three trials. Other significant peaks are not as robust. The peak near the 3-year period breaches the 99% confidence level in one of the three cases, and at least the 90% level in the other two. Peaks near the 3.5-year period, the 7- to 8-year period, and the 10- to 12-year period pass or nearly reach the 95% confidence level in two of the three cases, and at least the 90% confidence level in the remaining case.

Trend detection can also be accommodated through analysis of the LFV spectrum, but some subtleties must be taken into account. As explained in Section 3.1, multiple signals with period longer than N/p (where $p = 2$ is the bandwidth parameter used in all of our studies), corresponding to the secular band $f < pf_R = 0.02$ cycle/yr in this study, cannot be distinguished in the LFV spectrum. However, at least $K - 1 = 2$ distinct secular variations can still be separately identified by virtue of their orthogonal spatial patterns and temporal modulations if referenced to the secular band near $f = 0$. Both secular trends and ultra-low-frequency oscillatory signals in this case will be described as having a carrier frequency $f = 0$ and an envelope with $K - 1 = 2$ degrees of freedom (there is a loss of one degree of freedom at $f = 0$ due to removal of the mean). The envelope can thus describe limited oscillatory, though strictly not periodic, behavior. An iterative procedure is used to identify possible significant *secondary* secular modes of variation, based on the fraction of residual secular variance explained once the principal mode is accounted for. This process leads to the detection of two distinct secular time-scale signals at the 99% confidence level in the global temperature data set. The primary mode accounts for 77% of the zero-frequency variance (i.e., an LFV of 0.77) while the secondary mode describes most of the remaining 23% zero-frequency variance. Because the resolution of the LFV is variable between f_R and pf_R and $K - 1$ distinct modes (i.e., trends or oscillatory variations) are resolvable at each distinct frequency value, these two modes need not combine to describe *all* of the variance in the nominal secular band ($f < 0.02$ cycle/yr). Only for convenience are multiple secular variations referenced to the same frequency $f = 0$ for detection and reconstruction. A residual of secular band variance is left behind once these two secular variations are taken aside, describing the low-frequency noise background which is not discernible from colored noise.

Distinct peaks in the LFV spectrum at the 15- to 18-year, the 11- to 12-year, and the near-2.2-year period rise abruptly from the noise background, and are thus inferred as representing distinct "signals" in the data. In contrast, the group of peaks within the 3- to 7-year ENSO band are not well separated from the noise background or from each other, and may be associated with more complex frequency-domain structure than can be identified based on spectral analysis of a fixed window of data. In this case, the enumeration of distinct "signals" within the broader 3- to 7-year band seems inappropriate. Additional insights are obtained from evolutive generalizations of the procedure described in Sections 3.1 and 3.2, or further, from a wavelet-based multivariate decomposition (Park and Mann, 1998).

TABLE III STATISTICALLY SIGNIFICANT SPATIOTEMPORAL SIGNALS OR SIGNAL COMPONENTS ISOLATED IN THE SVD ANALYSIS, ENUMERATED IN ORDER OF INCREASING FREQUENCY, ALONG WITH ASSOCIATED RANGE IN FREQUENCY AND PERIOD OF THE SIGNAL / COMPONENT, LOCAL FRACTIONAL VARIANCE (LFV) EXPLAINED OF THE ASSOCIATED FREQUENCY BAND, MAXIMUM REGIONAL AMPLITUDE OF PATTERN, ROOT-MEAN-SQUARE AMPLITUDE OF PATTERN, AMPLITUDE OF GLOBAL MEAN OF PATTERN, AND PROJECTION OF PATTERN ONTO GLOBAL-MEAN TEMPERATURE[a] [FROM MANN AND PARK (1994).]

#	f (cycle/yr)	τ (years)	% LFV	T_{MAX}	T_{RMS}	T_{GLB}	P_{GLB}
Secular trend	0–0.02	trend	0.77	1.7	0.55	0.51	0.94
Multidecadal variation	0–0.02	> 50	0.23	1.4	0.29	0.03	0.10
Interdecadal oscillation	0.055–0.065	15–18	0.60	1.6	0.45	0.18	0.40
Quasidecadal oscillation	0.085–0.09	10–12	0.52	1.7	0.37	0.06	0.17
ENSO band	0.13–0.15	6.7–7.7	0.53	1.4	0.39	0.16	0.42
ENSO band	0.175–0.195	5.1–5.7	0.62	1.6	0.36	0.09	0.32
ENSO band	0.21–0.23	4.3–4.8	0.61	1.6	0.37	0.15	0.41
ENSO band	0.295–0.30	3.3–3.4	0.51	1.3	0.36	0.10	0.28
ENSO band	0.32–0.35	2.8–3.0	0.53	1.1	0.28	0.11	0.40
Quasibienniel oscillation	0.43–0.47	≈ 2.2	0.58	1.5	0.38	0.14	0.36

[a] For the latter statistic, $P_{GLB} = 1$, e.g., describes spatially uniform warming or cooling, while $P_{GLB} = 0$ describes complete spatial cancellation of regional variations in a global mean. Amplitudes of variability (in Celsius) correspond to maximum peak-to-peak cycle amplitude over the 100-year period.

It is nonetheless useful to examine separately the high- and low-frequency ENSO peaks, which we loosely refer to as ENSO "components" here. Table III itemizes the 10 significant peaks isolated in the MTM-SVD analysis.

As a crosscheck, we performed a parallel analysis using a small subset of ($M = 50$) gridpoints scattered evenly over the globe to estimate the effect of sampling inhomogeneity on the analysis of the global temperature data set. The resulting LFV spectrum (Fig. 19) is not significantly dissimilar from that of the full ($M = 449$) data set. The most notable differences are greater prominence of the 2- to 3-year time-scale ENSO peaks, and a slight shift in the location of the quasi-decadal peak (centered closer to the 10-year period in the sparse data set). We conclude that the MTM-SVD analysis of the global temperature data is reasonably robust to sampling inhomogeneity and variations in spatial sampling density. The bias introduced by the paucity of information in data-poor regions is more difficult

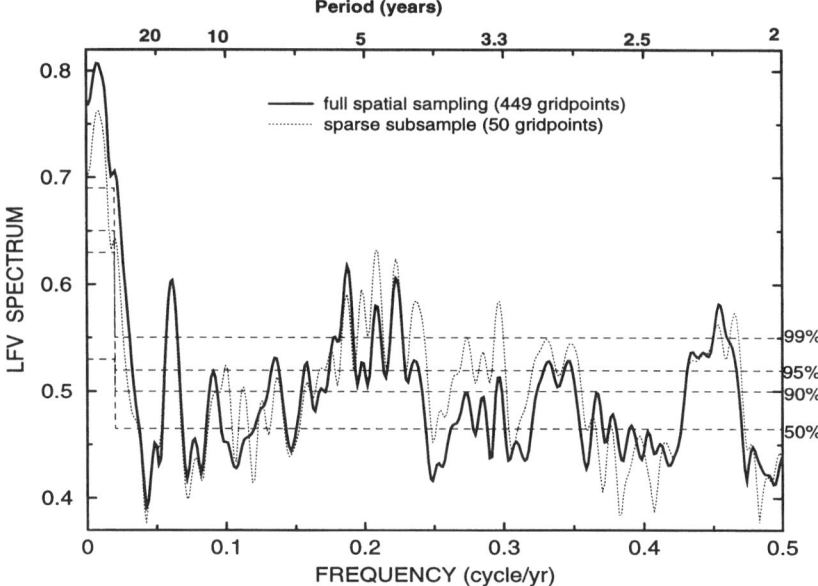

FIG. 19. LFV spectrum of the full $M = 449$ global temperature data set (solid) along with that of a sparse more homogeneously distributed $M' = 50$-gridpoint data set (dot-dashed). Horizontal dashed lines denote 90%, 95%, and 99% confidence limits from bootstrap resampling.

to determine. The exercises of Section 3.5 suggest, however, that these biases probably are not too influential for the spatial sampling available.

Spatial and Temporal Correlations across Time Scales

The similarity between the spatial patterns of distinct signals and the frequency components of a band-limited signal can provide insights into possible physical relationships between them. The squared dot-product of the spatial EOFs derived from the MTM-SVD analysis provides such a measure of similarity between spatial patterns. Under the assumption of Gaussian random spatial variations with $\tilde{M}$ complex spatial degrees of freedom, the statistical significances of such correlations are provided from standard tables. Table IV indicates statistical comparisons between the distinct signals or components identified. These quantitative comparisons of spatial relationships corroborate qualitative inspection of the temperature pattern reconstructions shown below. Patterns of temperature variability within the core 3- to 7-year ENSO band (i.e., those corresponding to peaks at 2.8- to 3.0-, 3.3- to 3.4-, 4.3- to 4.8-, and 5.1- to 5.7-year period) all

TABLE IV SQUARED CORRELATIONS BETWEEN SPATIAL AND TEMPORAL MODULATION PATTERNS OF DISTINCT SIGNAL / COMPONENT PAIRS WITH ASSOCIATED SIGNIFICANCE LEVELS[a] [FROM MANN AND PARK (1994).]

τ, years	trend	> 50	15-18	10-12	7-8	5.1-5.7	4.3-4.8	3.3-3.4	2.8-3.0	≈ 2.2
trend	1	0	0.16	0.13	0.44	0.33	0.25	0.19	0.10	0.77†
> 50	0	1	0.67	0.87†	0.11	0.51	0.49	0.54	0.30	0.21
15-18	0.13*	0.03	1	0.54	0.13	0.44	0.81†	0.95‡	0.52	0.15
10-12	0.00	0.06	0.37‡	1	0.25	0.55	0.23	0.38	0.40	0.44
7-8	0.13*	0.01	0.01	0.10	1	0.51	0.05	0.22	0.16	0.70*
5.1-5.7	0.04	0.04	0.25‡	0.22‡	0.03	1	0.20	0.27	0.04	0.70*
4.3-4.8	0.08	0.04	0.22‡	0.01	0.04	0.33‡	1	0.81†	0.25	0.04
3.3-3.4	0.07	0.02	0.25‡	0.16†	0.09	0.34‡	0.32‡	1	0.56	0.09
2.8-3.0	0.13*	0.01	0.11*	0.02	0.13*	0.25‡	0.42‡	0.25‡	1	0.07
≈ 2.2	0.02	0.01	0.05	0.26‡	0.17†	0.06	0.11*	0.19†	0.08	1

[a] The top triangle measures the similarity in the spectral EOFs or "principal modulations," comparing the long-term envelopes of different oscillatory signals or components and indicating possible temporal relationships. The bottom triangle compares spectral EOFs of different signals or components, measuring the similarity in their spatial patterns. Under the assumption of locally white noise, the significance levels for these correlations are estimated from the standard distribution of the spectral coherence based on $\tilde{M} = 20$ spatial and $K = 3$ spectral degrees of freedom, respectively. Symbols: *, > 90%; †, > 95%; ‡, > 99%.

share a characteristic global pattern (see, e.g., Halpert and Ropelewski, 1992) of in-phase variability throughout much of the tropics and extratropical teleconnection patterns such as the PNA or TNH pattern over the Pacific/North American sector (see Wallace and Gutzler, 1981; Barnston and Livezey, 1987). The corresponding spatial EOFs are correlated at greater than 99% confidence level (sharing between 25% and 50% of their variance in common in each case). The 7- to 8-year peak has a less classical ENSO pattern, but does show some significant similarity with the other ENSO-band patterns, as well as with other patterns (e.g., the quasi-biennial pattern) exhibiting a temperature signature consistent with the North Atlantic Oscillation (NAO—see, e.g., Wallace and Gutzler, 1981; Rogers, 1984; Barnston and Livezey, 1987; Lamb and Peppler, 1987). This observation seems to be consistent with the study of Rogers (1984) who observed a peak near the 7.3-year period in the spectrum of the NAO index. The spatial patterns associated with the interdecadal 15- to 18- and quasi-decadal 10- to 12-year signal both share a combination of ENSO-like and NAO-like features with certain (i.e., 3.3- to 3.4- and 5.1- to 5.7-year) ENSO components, exhibiting spatial correlations significant at > 99% confidence. Spatial patterns associated with quasi-biennial peaks centered near the 2.1- to 2.3-, the 7- to 8-, and the 11- to 12-year period are each dominated by an NAO-like pattern. While some of these correlations may be spurious, there are only 45 distinct correlation pairs, and very few should randomly exceed the 99% confidence level given our estimate of $\tilde{M} \geq 20$ spatial degrees of freedom in the data set.

In contrast, a similar dot-product of the spectral EOFs or "principal modulations" measures the similarity in the slow amplitude and phase modulation of distinct signals or components. Such similarity may be indicative of a nonlinear coupling between oscillatory variations with different periodicities, or simply modulation by the same long-term envelope. In the case of a secular mode and an oscillatory signal, a significant correlation may indicate modulation of the oscillatory signal by the secular variation. Under the assumption of a smoothly varying colored-noise background, the correlations should exhibit the null distribution describing pairs of Gaussian distributed complex K-vectors. These latter dot-products suggest a number of interesting interrelationships. The 15- to 18-year period interdecadal ENSO-like signal exhibits long-term temporal modulation similar to several of the ENSO-band components, suggesting the possibility of a consistent long-term modulation of ENSO during the past century. The secondary "multidecadal" secular variation exhibits a high correlation with the envelope of the 10-to 12-year quasi-decadal signal, suggesting the possibility that quasi-decadal oscillations in the North Atlantic are modulated by the long-term multidecadal secular variation centered in that region. In turn, there is some evidence for a long-term modulation of the quasi-biennial oscillation by the secular "trend." Such possible relationships are discussed later on.

Secular Signals

The primary secular mode (Fig. 20) describes an in-phase, global-scale secular trend in the temperature data set. The associated spatial pattern of warming projects strongly (94% projection) onto global-mean temperature.

The maximum regional warming trend (Table III) is ~ 1.7°C, while the estimated global-mean warming is closer to 0.5°C. Both this mean warming and the temporal reconstruction of the trend (inset, Fig. 20) are consistent with recent nonparametric analyses of global-mean temperature (Ghil and Vautard, 1991), indicating warming most rapid from about 1920 to 1950. The continued acceleration of global warming in certain regions through the 1990s is not associated, in the context of our study, with long-term secular warming. Such warming could represent a sudden, nonstationary shift to sustained ENSO-like global climate patterns (see Trenberth and Hoar, 1995), but may also be consistent with the natural internal fluctuations of the tropical Pacific system (Cane *et al*., 1997). This is highlighted

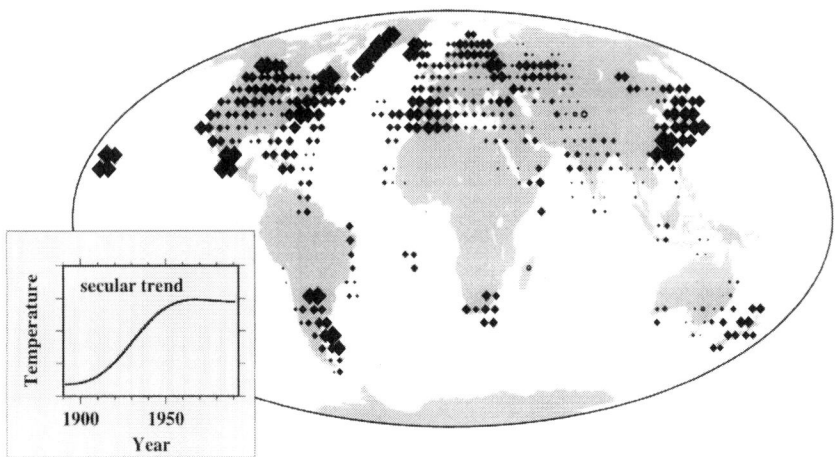

FIG. 20. Pattern of temperature variability associated with the primary secular mode. Solid diamonds are used to indicate gridpoints evolving positively with the reference time-domain signal shown in the inset (warming), while open circles represent the few gridpoints evolving negatively with the time-domain signal shown (cooling). As in all subsequent plots, the symbol sizes scale the relative magnitude of temperature variations. The absolute scale for typical and maximum regional variations is provided by Table III. [From Mann and Park (1994).]

by the fact that a marked positive recent departure from the low-frequency global warming trend is found in the far eastern tropical Pacific (within the reach of positive El Niño SST anomalies), but not in the more central subtropical Pacific (see Section 4.2, Fig. 30). Much of the warmth of the 1980s (but *not* the 1990s) is, in contrast, explained by interdecadal fluctuations (see below).

Though associated with a clear global warming signal, the spatial pattern is variable, with certain gridpoints actually cooling slightly. Such regional departures from the average warming of ~ 0.5°C suggest the probable existence of associated atmospheric circulation anomalies accompanying the warming signal. Such circulation anomalies are addressed more directly in the joint temperature/SLP analysis of Section 4.2. The strongest warming (~ 1.7°C) is observed along the margin of Greenland, which could indicate the influence of a positive ice-albedo feedback. Further such evidence is provided by the changes in seasonality of temperature described by Mann and Park (1996a). For comparison, a calculation of the simple linear trend in the gridpoint data yields a similar pattern, albeit with a moderately higher estimated warming signal (compare the resulting values $T_{MAX} = 1.87°$, mean region warming $T_{RMS} = 0.59°$, and average global warming $T_{GLB} = 0.57°$ with their corresponding values in Table III), owing largely to the statistical leverage of recent decadal-scale warming in a nonrobust least-squares trend estimation.

In contrast to the primary mode, the significant secondary secular mode describes a spatially heterogeneous multidecadal "cycle" centered largely in the North Atlantic (Fig. 21) which projects little (10%) onto global-mean temperature. To some extent the dissimilarity in the features of the two secular modes is guaranteed by their mutual orthogonality, much as are the modes of a classical time-domain PCA. The principal mode tends to capture the in-phase global component, while the secondary mode favors the dominant mode of variation in the spatially heterogeneous residual (see, e.g., the synthetic PCA experiment of Section 2.2). Some linear combination of the two modes (i.e., rotation) or other statistical decompositions could possibly provide a more physical decomposition. Schlesinger and Ramankutty (1994), for example, identified a residual multidecadal oscillatory component in the global temperature record after removing a model-predicted trend due to anthropogenic forcing. Only analyses of longer data sets (e.g., proxy data—see Section 4.3) can more objectively separate out possible anthropogenic trends and low-frequency oscillatory features of the climate.

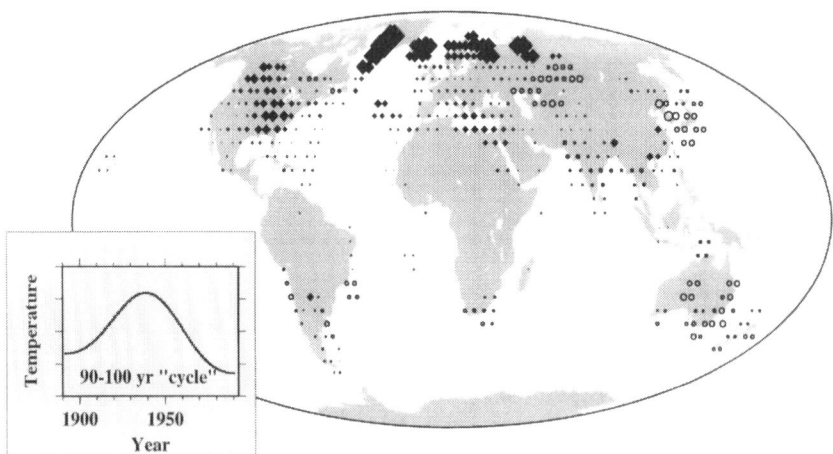

FIG. 21. Pattern of temperature variability associated with the secondary secular mode of variation. High-amplitude variability is confined largely to the North Atlantic. Each gridpoint evolves with the same 90- to 100-year "oscillation," shown in the inset, differing only in magnitude or sign. Solid diamonds evolve in phase with the North Atlantic, while gridpoints with open circles vary oppositely. A minimum slope constraint was invoked for the temporal signal reconstruction. [From Mann and Park (1994).]

The cycle-like characteristic of the secondary secular variation is poorly constrained by the short duration of the data set and the simple boundary constraints invoked. However, a more objective signal-reconstruction approach based on the optimal weighting of various boundary constraints (see Section 4.2) favors a single ~ 70-year multidecadal cycle of variation, and long-term proxy data (Section 4.3) offer additional evidence for a multidecadal or "century-scale" oscillatory mode with a 50- to 70-year time scale. The pattern of this multidecadal variation (Fig. 21) exhibits significant amplitude in the high-latitude North Atlantic (as large as 1.4°C; see Table III), in phase with smaller amplitude variability in the United States, Northern Europe, and the Mediterranean region, and out of phase with variability elsewhere over the globe. The cycle of warming in the North Atlantic from roughly 1890–1940, and subsequent cooling, is consistent with the long-term trend in North Atlantic sea surface temperature and air temperatures determined elsewhere (Deser and Blackmon, 1993; Kushnir, 1994; Schlesinger and Ramankutty, 1994). Large temperature variations in the high-latitude North Atlantic support a possible connection with century-scale variability in deep water production (e.g., Stocker *et al.*, 1992). The opposite sign of anomalies in the North and South Atlantic, consistent with the interhemispheric contrast pattern noted by Folland *et al.* (1986), is consistent with changes in cross-equatorial heat flux that would be expected to arise from variability in the thermohaline circulation. The near-cancellation of the pattern in a global average suggests a process that largely redistributes heat over the Earth's surface. Additional dynamical insight is offered by the analysis of associated atmospheric circulation variations in Section 4.2, while better statistical constraint on the apparent multidecadal oscillation is provided by the multiproxy data analysis of Section 4.3.

Interdecadal Signal

The interdecadal 15- to 18-year-period signal shown in Figure 22 has a pattern consistent with that of the pair-coherence map of Mann and Park (1993) and probably corresponds to the bidecadal temperature signal identified by Ghil and Vautard (1991) in global-average temperature. The spatial pattern resembles, at least superficially, that of ENSO, with high-amplitude, largely in-phase tropical warming or cooling, and similar inferred extratropical teleconnection patterns. Despite significant cancellation in the global average, this mode is associated with a sizable peak-to-peak T_{GLB} ~ 0.2°C, with a maximum global warm anomaly coincident with tropical warming. The time-domain signal (inset of Fig. 22) suggests that the anomalous warmth of the 1980s was associated, at least in part, with a

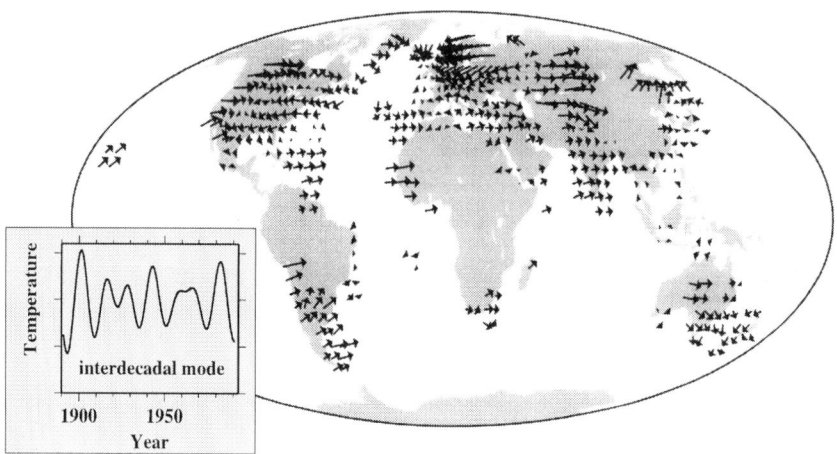

FIG. 22. Pattern associated with the 15- to 18-year interdecadal mode, with zero phase (e.g., tropics) evolving with the time-domain signal shown in the inset. Maximum pattern average warmth is coincident with tropical warmth. The size of the vectors indicates relative magnitude of temperature variations (absolute scale is provided in Table IV). Vector orientation indicates the relative temporal lag at each gridpoint. We define a zero phase vector (i.e., gridpoints with a vector pointing rightward, toward "3 o'clock," for example, much of the tropics in this case) as evolving positively with the time-domain signal shown in the inset. Counterclockwise rotation in the spatial pattern indicates positive lag, while clockwise rotation represents negative relative lag or "lead." A rotation of 360° corresponds to the periodicity of the mode ($\sim$ 17 years in this case). All gridpoints share the same long-term modulation, with the $\sim$ 17-year carrier oscillation shifts forward or backward with the phase lags indicated in the spatial pattern. For example, gridpoints with vectors at "12 o'clock" experience maximum warming at a 90° $\sim$ 4-year lag relative to peak tropical warming. Gridpoints at "6 o'clock" experience maximum warming at $\sim$ 4 years before peak tropical warming, and gridpoints at "9 o'clock" experience maximum cooling simultaneous with peak tropical warming. The pattern average variability is nearly in phase with the tropical variability, so that peak projection onto global warmth corresponds to peak tropical warmth. [From Mann and Park (1994).]

large positive excursion of the interdecadal oscillation. Note, however, that global warming in the interdecadal oscillation is associated with simultaneous cooling in the southeastern United States and Europe.

In addition to exhibiting spatial correlations at $>$ 99% confidence with three of the four ENSO spatial patterns (Table IV), the interdecadal mode appears to be associated with a modulation common to two of the ENSO-band components. The signal thus appears to exhibit at least a limited connection with long-term variations in ENSO, as has been suggested in previous studies of interdecadal climate variability (Trenberth, 1990; Tanimoto et al., 1993). This signal has been independently observed in studies of the South Atlantic (Venegas *et al.*, 1996), and may also be related to significant regional climate impacts such as influences on Sahel rainfall (Folland *et al.*, 1986) and the dynamic topography of the oceans (Unal and Ghil, 1995).

A variety of mechanisms have been offered to explain such interdecadal ~ 15- to 30-year oscillatory behavior of the climate, including external astronomical forcing (e.g., Royer, 1993), high-latitude ocean–atmosphere interactions (e.g., Darby and Mysak, 1993; Mysak and Power 1992), nonlinear instabilities in the global thermohaline circulation (e.g., Chen and Ghil, 1995) and extratropical ocean–atmosphere feedback mechanisms (e.g., Trenberth and Hurrell, 1994; Latif and Barnett, 1994—henceforth LB94; Von Storch, 1994), intrinsic tropical mechanisms (Graham, 1994), and coupled tropical/extratropical ocean–atmosphere mechanisms (Gu and Philander, 1997) in the Pacific ocean. Possible mechanistic explanations of the observed variability are further explored in Section 4.2.

Quasi-Decadal Signal

A "quasi-decadal" signal with 10- to 12-year periodicity (Fig. 23) exhibits a temperature pattern with large regional amplitude in the North Atlantic, exhibiting a quadrupole phase pattern of warming and cooling in different regions bordering the Atlantic basin. This pattern is reminiscent of an

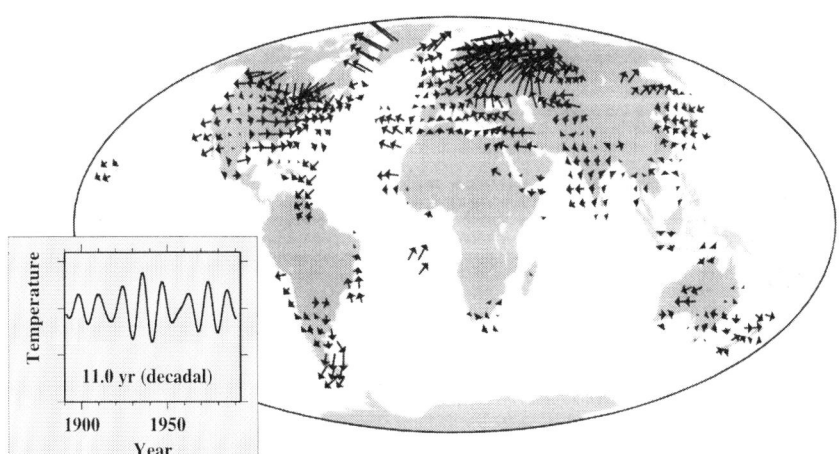

FIG. 23. Spatial pattern of variability associated with the 10- to 12-year decadal mode, with the zero phase signal (i.e., arrow pointing directly right—e.g., Great Britain, most of southeastern United States) corresponding to the time series shown in the inset. Symbol conventions are similar to Figure 22. [From Mann and Park (1994).]

NAO influence. Such a quasi-decadal signal has independently been observed in other studies of North Atlantic climate data (Deser and Blackmon, 1993; Hurrell, 1995). The surface temperature pattern tends to cancel, however, in a large-scale average, exhibiting a relatively weak projection (18%) onto global-mean temperature. It is thus not surprising that such a signal is either undetected (Folland et al., 1984; Ghil and Vautard, 1991) or weakly detected (Allen and Smith, 1994; Mann and Lees, 1996) in studies of global-mean temperature, but emerges more distinctly in regional studies of continental United States temperature (Mann et al., 1995a; Dettinger et al., 1995), SLP (Mann et al., 1995a), precipitation (Currie and O'Brien, 1992), Indian precipitation (Vines, 1986; Mitra et al., 1991), and both tropical (Houghton and Tourre, 1992; Mehta and Delworth, 1995; Chang et al., 1997) and North Atlantic (Deser and Blackmon, 1993) regional climate data. There is some debate in the literature over whether this decadal signal is or is not characterized by a cross-equatorial dipole in Atlantic sea surface temperatures (Houghton and Tourre, 1992; Mann and Park, 1994; Mehta and Delworth, 1995; Chang et al., 1997; Tourre et al., 1998). Where there is spatial sampling in the Atlantic, our pattern exhibits fairly clear evidence for a nearly 180° reversal in phase of surface temperature variations north and south of the equator, supporting Chang et al. (1997) and Tourre et al. (1998). The significant relationship (95% significance level—see Table IV) between the long-term modulation of this quasi-decadal oscillation and the multidecadal secular variation in the North Atlantic described above, suggests that the longer multidecadal cycle of warming and cooling in the North Atlantic may modulate the amplitude of these higher-frequency quasi-decadal oscillations. In contrast, the signal is not significantly correlated ($< 50\%$ confidence level) with the $\sim$ 11-year sunspot cycle, casting doubt on a simple linear relationship between decadal-scale solar forcing and similar-time-scale surface climate variations (e.g., Currie and O'Brien, 1992). Joint analysis of historical sea-level pressure and surface temperature, discussed in Section 4.2, suggests possible mechanisms for the quasi-decadal oscillation.

ENSO Signal

The MTM-SVD analysis of the global temperature data reveals clear ENSO-band signals (Fig. 24), and confirms a rough division of the ENSO-related variability into higher-frequency (2.8- to 3.0- and 3.3- to 3.4-year period) and lower-frequency (4.3- to 4.8- and 5.1- to 5.7-year period) bands, consistent with the notion of distinct high- and low-frequency bands of ENSO (e.g., Keppenne and Ghil, 1992; Dickey et al., 1992). A more

(a)
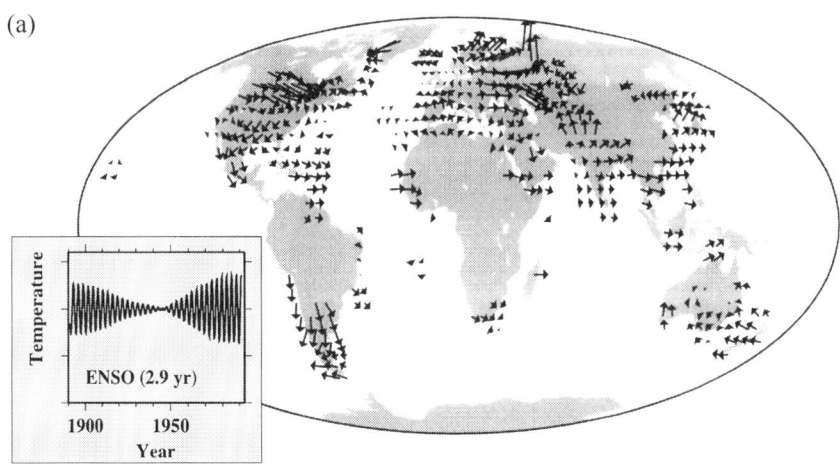

(b)
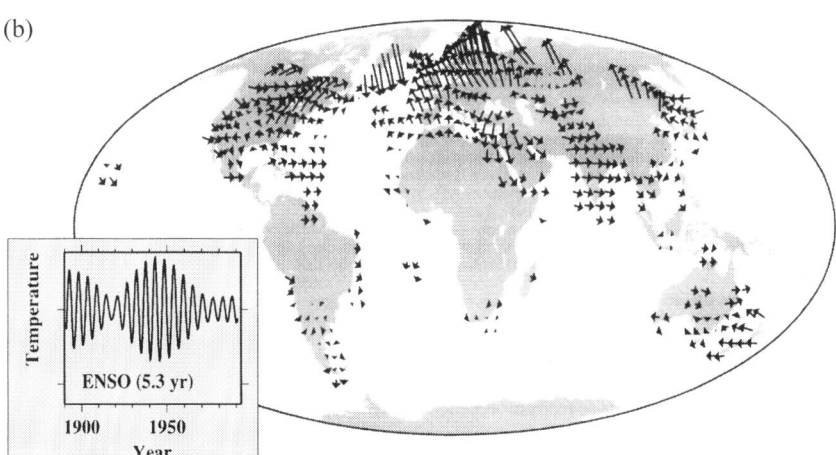

FIG. 24. Spatial pattern associated with the 2.8- to 3.0-year (a) and 5.7- to 5.7-year (b) ENSO components. Symbol conventions are similar to Figure 22, with zero phase (e.g., much of the tropics) coincident with the time-domain signal shown in the inset for both cases. [From Mann and Park (1994).]

elaborate time-dependent description of the frequency-domain characteristics of ENSO is provided in the analysis of Section 4.2. The spatial patterns (Fig. 24) associated with the ENSO spectral peaks are similar to the ENSO temperature pattern identified by Halpert and Ropelewski [(1992), henceforth HR92]. For example, both HR92 and our 2.8- to 3.0-year ENSO component show a pattern of roughly in-phase tropical warming coincident with warming and cooling, respectively, in the northwest and southeast United States. Such behavior is consistent with the positive phase of a PNA or TNH circulation anomaly pattern that has been argued to be favored during tropical warm events (Horel and Wallace, 1981; Livezey and Mo, 1987). Both patterns also share warming in eastern Europe coincident with cooling in central Asia and slight cooling in western Europe, as well as cooling in the northern Pacific that is in phase with warming in the tropical western Pacific. This latter pattern resembles a similar Western Pacific Oscillation or "WPO" pattern. Though much of warming during ENSO is produced by tropical ocean–atmosphere heat exchange, some of this heat is transported poleward by a variety of processes. Hence there is a tendency for warming, for example, in middle as well as tropical latitudes during or shortly following warm events, even though anomalous advection can produce cooling in certain regions. Such dynamics are responsible for the nontrivial impact of ENSO events on global-mean temperature. Each of the ENSO patterns has significant projections onto global temperature (Table III), with global-average peak-to-peak fluctuations $T_{GLB} \sim 0.10°C$ for each component, and somewhat larger excursions for the sum over all components. Other workers (Jones, 1989; Angell, 1990) have noted similar interannual variations in global-average temperature associated with ENSO.

Each of the ENSO component patterns shares features of in-phase tropical warming with extratropical patterns consistent with combinations of the WPO, PNA, and TNH patterns. The lowest-frequency (5.1- to 5.7-year ENSO) component exhibits the most prominent NAO signature, consistent with Rogers (1984), who identified a peak near 6-year period in the cospectrum of the Southern Oscillation and NAO. While the pattern for each component is by definition linearly reversed for negative and positive excursions, the sum of the components which interfere constructively or destructively to describe the actual "warm" and "cold" events (see, e.g., the total signal reconstructions shown later) need not exhibit an equivalence between large negative and positive reconstructed events. In each of our ENSO patterns, maximum global warm anomalies coincide with tropical warm events, while the maximum cold anomalies coincide with cold events.

Quasi-Biennial Signal

A quasi-biennial (~ 2.2-year-period) signal is isolated in the global temperature data (Fig. 25) consistent with the independent detection of such a signal in Northern Hemisphere SLP (Trenberth and Shin, 1984), North Atlantic SLP, winds, sea-level pressures (Deser and Blackmon, 1993) and air temperatures (Gordon *et al.*, 1992), and predominantly north−south variations in U.S. temperature (Dettinger *et al.*, 1995). The signal exhibits a prominent NAO temperature pattern as well as other extratropical regional anomalies, and is associated with sizable regional variations (0.38°C mean regional amplitude variations), as well as a significant peak-to-peak global-average variation (0.14°C). As discussed by Trenberth and Shin (1984) and below in Section 4.2, the surface quasi-biennial signal shows no obvious relationship with the stratospheric quasi-biennial oscillation (QBO) of similar time scale, although such connections have been suggested with atmospheric data higher in the troposphere (Labitzke and van Loon, 1988) or indirectly, through modulation of ENSO (Barnston *et al.*, 1991). Interestingly, there is evidence of modulation by the same envelope which modulates the lower-frequency ENSO variability (precisely those ENSO-band components, the 5.1- to 5.7-year and 7- to 8-year components which exhibit a notable NAO pattern—Table IV). These correlations are only marginally (~ 90%) significant, however, and may well be spurious.

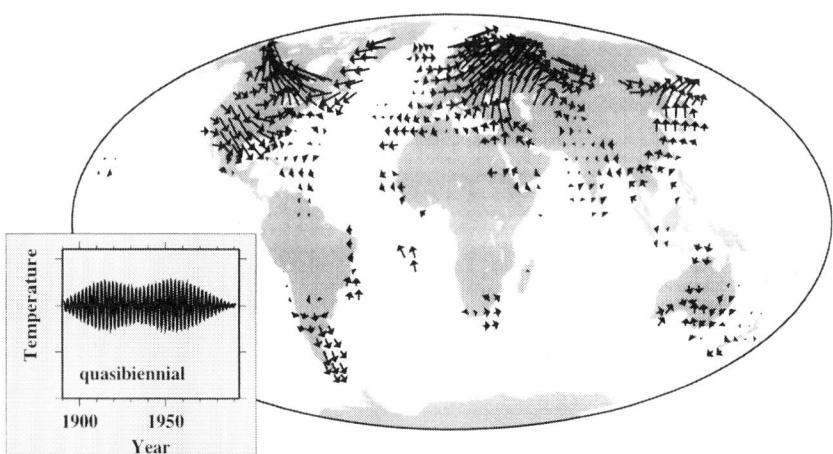

FIG. 25. Pattern of variability associated with the 2.1- to 2.3-year quasi-biennial mode. Zero phase variability (e.g., England) evolves positively with the time series shown in the inset. Symbol conventions are similar to Figure 22. [From Mann and Park (1994).]

Single-Gridpoint Reconstructions

Figure 26 shows the time-domain signal reconstructions for several chosen gridpoints, based on the sum over each of the spatiotemporal signals or signal components isolated in the MTM-SVD analysis. Also shown are the interannual lowpassed (periods longer than 2 years retained) raw data for comparison. The North American east coast gridpoint (Fig. 26a) is dominated by the secular warming trend (18% of the lowpassed "interannual" data variance) and interannual (ENSO and quasibiennial) fluctuations ($\approx 13\%$ of the interannual variance), while the quasi-decadal signal describes about 2% of the interannual variance. The total reconstruction describes 34% of the variance. The Hawaiian gridpoint (Fig. 26b) is dominated by the secular warming signal (61% of the interannual variance), but with a large share of variance also described by the interdecadal signal ($\approx 8\%$). Seventy-four percent of the total lowpassed variance is explained by the reconstructed signal. Finally, the South American Pacific-coast gridpoint (Fig. 26c) is in a region where the direct influence of El Niño can be expected. The reconstruction, not surprisingly, is dominated by the interannual ENSO-band signals ($\approx 20\%$ of the interannual variance). All strong El Niño events (e.g., Quinn and Neal, 1992) are clearly captured in the reconstruction. Significant variance is also described by both secular variations ($\approx 5\%$ and 4%) and the interdecadal signal ($\approx 6\%$). The total reconstructed signal describes 35% of the total interannual variance.

The interannual variance explained by the total signal reconstructions range over $\sim 1\%$ to $\sim 75\%$ for the $M = 449$ gridpoints in the analysis, describing $\sim 40\%$ of the total interannual variance in the multivariate data set. Sixteen percent of the total variance is accounted for by the two secular variations, so that the interannual- and decadal-scale signals describe an important share of the low-frequency multivariate data variance. The variance not explained is attributed in our paradigm to a smoothly varying colored-noise background. However, some of that variance may be attributed to episodic or event-like signals which do not exhibit clear frequency-domain structure (e.g., the impacts of explosive volcanism on surface temperatures—e.g., Bradley and Jones, 1993), nonlinear or chaotic features of variability (e.g., Lorenz, 1990) which are not captured well by a

FIG. 26. Specific gridpoint time-domain signal reconstructions for chosen regions determined from summing over each of the spatiotemporal signals or components isolated in the analysis. The reconstructed signals are shown (solid curves) along with the interannual lowpass of the raw data (dotted curves). (a) North American East Coast Gridpoint (containing New York City). (b) Tropical Pacific Gridpoint (containing main island of Hawaii). (c) South American Grid point containing Santiago, Chile. [From Mann and Park (1994).]

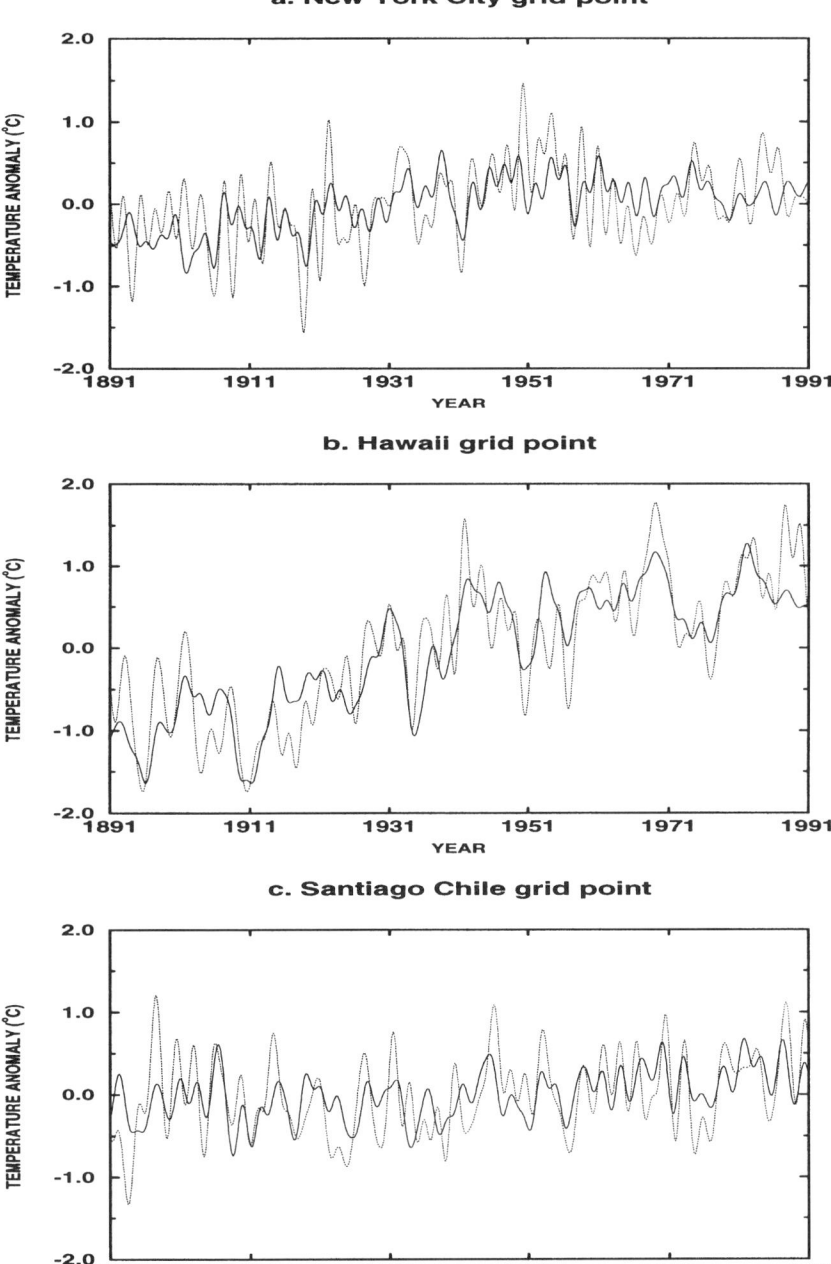

linear frequency-domain decomposition, and in the present analysis the imposition of a priori (and hence, not in general optimal) boundary constraints in temporal signal reconstruction from frequency-domain information. An attempt is made to improve upon the latter limitations in the analysis described in the following section.

Summary

Using 100 years of global temperature anomaly data, the MTM-SVD method isolates coherent spatiotemporal oscillations of global climate variability. Organized interannual variability appears to be associated either with ENSO, or with extratropical patterns that chiefly involve an NAO pattern. Secular variance is dominated by a globally coherent trend, with nearly all gridpoints warming in phase at varying amplitude. A smaller, but significant, share of the secular variance corresponds to a pattern dominated by warming and subsequent cooling in the high-latitude North Atlantic with a roughly centennial time scale. Spatial patterns associated with significant peaks in variance within a broad period range from 2.8 to 5.7 years exhibit characteristic ENSO patterns. A recent transition to a regime of higher ENSO frequency is suggested by our analysis. An interdecadal mode in the 15- to 18-year-period range appears to represent long-term ENSO variability. This mode has a sizable projection onto global-average temperature, and accounts for much of the anomalous global warmth of the 1980s. A quasi-biennial mode centered near a 2.2-year period and a mode centered at a 7- to 8-year period both exhibit predominantly a North Atlantic Oscillation (NAO) temperature pattern. A "decadal" mode centered on an 11- to 12-year period also exhibits an NAO temperature pattern, and may be modulated by the century-scale North Atlantic variability. Decadal variability has weak impact on global-average temperature, but gives rise to a strong redistribution of surface heat.

4.2. Northern Hemisphere Joint Surface Temperature and Sea-Level Pressure Data

The potential dynamical insights possible from analyzing the joint relationship between surface temperature and atmospheric circulation are well established (e.g., Namias, 1983; Cayan, 1992a, b). Here we investigate spatiotemporal signals in joint fields of surface temperature and SLP in the Northern Hemisphere which provide near-uniform coverage from the tropics through the subpolar regions for almost a century. A variety of dynamical inferences are possible in the case of the joint-field analysis, complementing the observations from the global temperature analysis described in the previous section. Simple relationships between SLP anomaly patterns, inferred surface circulation anomalies, and associated relative advective effects (warming or cooling) may suggest a largely passive response of the temperature field to circulation anomalies. Cold-season warming/cooling along coastlines associated with anomalous inferred onshore/offshore circulation suggests variability in the degree of continental vs maritime influence. SST anomalies that do not reflect a passive response to atmospheric circulation anomalies may indicate underlying changes in oceanic circulation and ocean–atmosphere exchange. An analysis of the signatures provided by the joint spatial patterns thus provides insight into the underlying mechanisms of organized climate variations. Comparison with recent coupled ocean–atmosphere model simulations allows us to determine if, beyond empirical statistical inferences, there is some physical motivation for placing confidence in an apparent climatic signal.

The data used consist of the land air and sea surface temperature anomaly data set described and used in Section 4.1 (Jones and Briffa, 1992; Jones, 1994), and gridded sea-level pressure data on a similar but staggered $5° \times 5°$ grid in the Northern Hemisphere (Trenberth and Paolino, 1980—this data set is continually updated through the NCAR archive). We confine our analysis to the latitude band $17.5°-72.5°N$, allowing fairly thorough spatial coverage for both fields. We use the 95-year ($N = 1140$ months) interval 1899–1993 for which both SLP and temperature data are available. Only gridpoints with nearly continuous monthly sampling (very few gaps, and no single gap longer than 12 months) are used, gaps filled as earlier by simple interpolation. This criterion for selection yields a set of $M = 601$ temperature gridpoints and $P = 792$ SLP gridpoints, nearly covering the subtropical-to-subpolar region of the Northern Hemisphere (Fig. 27), with some relatively modest spatial gaps in the temperature data set.

Possible sources of bias in the temperature data were discussed in Section 4.1. The gridded SLP data exhibit potential biases of their own. The SLP field for any given month represents the spatial interpolation of often sparse observations by hand-drawn analyses. These data are of somewhat questionable quality during the earliest (pre-1922) part of the century and during World War II. See Trenberth and Paolino, (1980) for a detailed discussion of the quality and potential sources of bias in this data. From a spatial point of view, the data quality is poorest in high-altitude regions with strong cold-season inversions (e.g., large parts of Asia) where sea-level reductions of surface measurements may be flawed. Thus, while there is potentially useful information in these data throughout the twentieth century, inferences that depend heavily on the behavior of the data

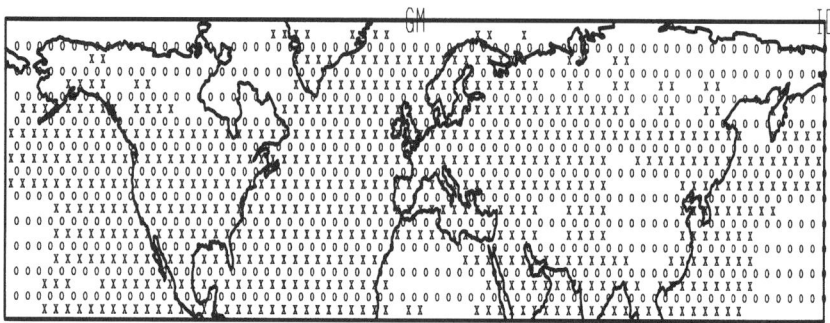

FIG. 27. Locations of gridpoints used in analysis, with locations of temperature data indicated by "x"s, and SLP data indicated by "o"s. The grid has latitudinal extent 15°–70°N and longitudinal extent −180E to 180E, with SLP gridpoints located every 5° and temperature gridpoints staggered 2.5°. The positions of the Greenwich meridian and the international dateline, as in following plots, is shown for reference. [From Mann and Park (1996b).]

early in the period (such as the spatial details of long-term trends) must be caveated by consideration of the potential data quality problems. Many potential sources of bias, however, are diminished in influence by our seeking organized, dynamically consistent variations in both temperature and SLP fields. Errors that are isolated in time, for example (e.g., data problems associated with World War II), should not have a large influence on our signal-detection or reconstruction procedure.

Here we apply a modified version of the standard MTM-SVD analysis described in Section 3.1, with the conventional choices $K = 3$ spectral degrees of freedom and bandwidth parameter $p = 2$, to the 95 years of overlapping monthly data for both SLP and temperature data fields. The basic MTM-SVD procedure is modified to determine the dominant modes of narrowband variability in two fields—surface temperature and SLP—simultaneously. This joint-mode analysis is accomplished by application of the MTM-SVD analysis to the aggregate data matrix for the two fields, and is a frequency-domain analog of the combined principal component analysis or CPCA approach described by Bretherton *et al.* (1992) and applied by Wallace *et al.* (1992) to the eigendecomposition of joint atmospheric data fields. In the joint-field MTM-SVD analysis, each constituent series is standardized for the analysis. To account for the small differences in the sizes of the two data sets ($M = 601$ temperature gridpoints vs $P = 792$ SLP gridpoints), the weights on the corresponding entries in the data matrix are adjusted so that the two fields contribute equal total standardized variance. Temporal signal reconstruction is performed based on the objective boundary constraints described in Section 3.2. Signals for which time-evolving frequency structure is detected are reconstructed with the evolutive signal-reconstruction technique (Section 3.2).

It is appropriate here to note some possible caveats regarding multivariate joint-field decompositions. In the context of conventional time-domain PCA decomposition, the relative strengths and weaknesses of alternative methods of decompositions for joint or "coupled" fields are explored in some detail by Bretherton *et al.* (1992). Potential deficiencies and limitations of joint-field generalizations on time-domain PCA (which are unfortunately sometimes referred to misleadingly as simply "SVD") have been pointed out recently by a number of authors (e.g., Newman and Sardeshmukh, 1995; Cherry, 1997). The primary limitation is that the joint-field patterns obtained are not robust, depending quite sensitively, for example, on the relative variance contributions of the two fields. This sensitivity can be understood in terms of the inherent ambiguities in specifying an objective rotation of EOFs in conventional PCA. (A nice explanation of the relationship between joint-field PCA and rotated PCA analysis is provided by Cherry (1997).) In fact, generalizations related to varimax rotation have been suggested as a possible means of specifying more objective joint-pattern decompositions (Cheng and Dunkerton, 1995). Nonetheless, much as MTM-SVD largely removes the ambiguity associated with rotation (recall the discussion of Section 3.4), the narrowband decomposition specified by MTM-SVD projects joint spatial patterns which are quite robust, being widely insensitive to variations in the relative contributions of the two independent fields. This point is discussed further below in the context of application to the joint SLP/surface temperature data set.

LFV Spectra

The LFV spectrum for the joint-field analysis (Fig. 28) is consistent with that obtained in the analysis of global temperature data described in Section 4.1 (see Fig. 18), with significant variance peaks at the quasi-biennial (2.1- to 2.2-year) time scale, within the 3-to 7-year ENSO-period band, and at "quasi-decadal" (10-to 11-year-period) and "interdecadal" (16- to 18-year-period) time scales. Two significant modes are also identified within the secular band corresponding to variability on time scales $\tau > 48$ years in the 95-year data set. It is worth noting that the quasi-decadal and interdecadal peaks exhibit a less distinct signal separation than was evident in the corresponding LFV spectrum of the global temperature data; this is not especially surprising given that the differing spatial signatures of the two signals were most evident (compare Figs. 22 and 23)

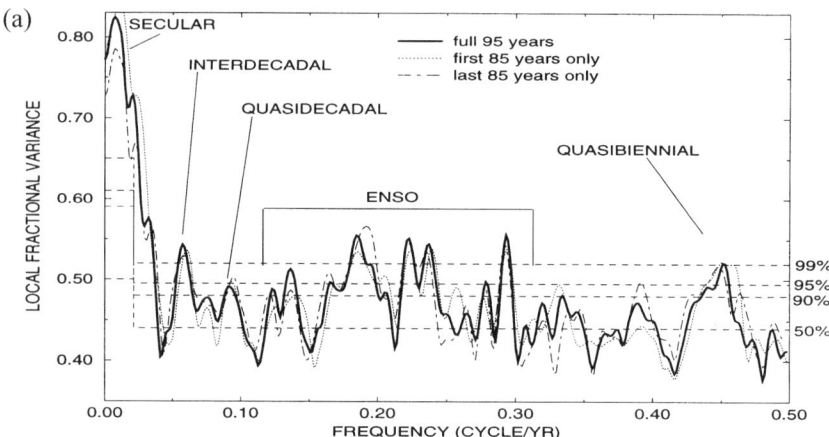

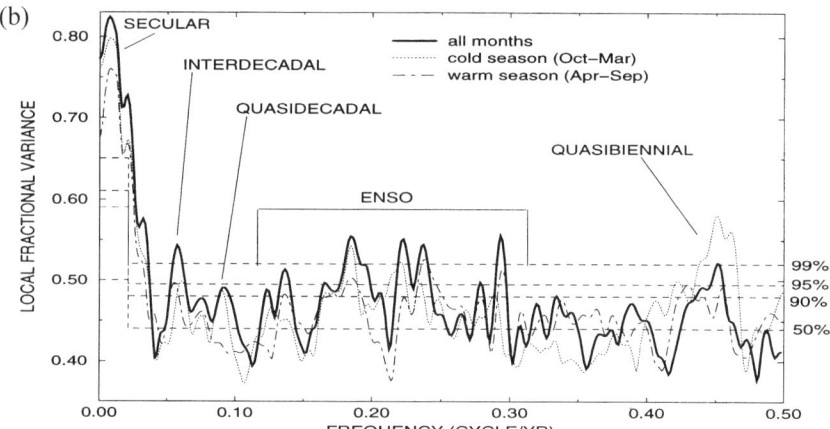

FIG. 28. (a) LFV spectrum for the 1140-month (95-year) joint surface temperature and SLP data set. Horizontal dashed lines denote median (50%), 90%, 95%, and 99% confidence limits from bootstrap resampling. (b) Comparison of LFV spectrum for all-months and cold- and warm-season analyses. The numerical vertical scale and significance levels shown apply to the all-months analyses. The LFV spectra for the cold- and warm-season-only analyses are slightly renormalized (by the factors ≈ 1.02 and 0.98, respectively) so that the quantiles of the null distribution and significance levels shown apply roughly for each case, even though the effective number of degrees of freedom in the sample varies between the two distinct seasonal windows. [From Mann and Park (1996b).]

in the tropical and southern latitudes not present in the sampling of the joint data fields analyzed here. A restricted sensitivity test (see Fig. 28a) using 85-year data subsegments long enough to resolve the secular and relatively closely spaced quasi-decadal and interdecadal peaks demonstrates relative stability in the significance of the peaks detected in the analysis of the full 95-year data set. For the interannual signals, a more liberal sensitivity test or "evolutive analysis" can be performed, as described below. The LFV spectrum was further found to be similar whether the MTM-SVD decomposition was performed on the two fields (SLP and temperature) with equal contributions of variance, or performed on the temperature data alone with the resulting signals linearly projected onto the SLP fields. Somewhat different results were obtained when the analysis was performed on the SLP field alone, which we attribute to the questionable quality of the SLP data (especially early in the twentieth century), which limits their usefulness by themselves for signal detection and reconstruction.

Comparison of LFV spectra for all-seasons, warm-, and cold-season data (Fig. 28b) suggests that secular and interdecadal signals are seasonally robust, although the spatial signatures of these signals, mostly in mid and high latitudes, are shown below to display seasonally specific features. The interdecadal signal appears to derive its strong ($> 99\%$) significance in the all-seasons analysis from more moderate amplitude, but consistently significant expression, during both seasons. In contrast, the quasi-decadal signal appears as a distinctly cold-season phenomenon, as prominent as the interdecadal signal during that season but completely absent in the warm season. The ENSO-band variance peaks show some distinct differences between the all-seasons and seasonal analyses. The quasi-biennial signal is significant during both independent seasons, but is clearly stronger during the cold season (we have checked, through analysis of synthetic examples, that no significant bias is introduced by the proximity of quasi-biennial frequencies to the $f_N = 0.5$ cycle/yr Nyquist frequency in the case of these seasonal analyses). It is possible that a seasonally persistent signal is only *detectable* above noise at certain times in our analysis. For example, a low-frequency carrier signal in tropical SST may only lead to a larger-scale expression during the winter season when high-amplitude extratropical circulation anomalies are induced by associated tropical heating anomalies. Since tropical coverage in our spatial sampling is marginal, such a signal may not be detectable during the warm season, as the associated tropical carrier signal would be missed. Such limitations should be kept in mind. An independent analysis of the spatial and temporal patterns for both seasons provides insight into such issues. An evolutive LFV spectrum (Fig. 29) was calculated using an $N = 40$-year moving window through the data set. This choice of window width admits a frequency resolution 0.10 cycle/yr so that the quasi-decadal and interdecadal peaks of Figure 28 are not distinguishable, and only periods shorter than 20 years can be confidently separated from secular variations. The relative stability of interannual signals, and the potential time-dependence of the amplitude and frequency characteristics, can, however, be tested. A 40-year window allows roughly one phase discontinuity per 13 years. Thus, low-frequency (5- to 7-year time scale) ENSO variability is assumed to maintain coherence (though varying in amplitude) over 2–3 "cycles." In contrast, the quasi-biennial oscillation is statistically modeled as being phase-coherent (and varying slowly in amplitude) over roughly 5 cycles with a 40-year window, which may imply a physically unrealistic time scale of coherence. For this signal we employed (see below) a somewhat shorter 20-year window in evolutive time-reconstruction of the signal. A

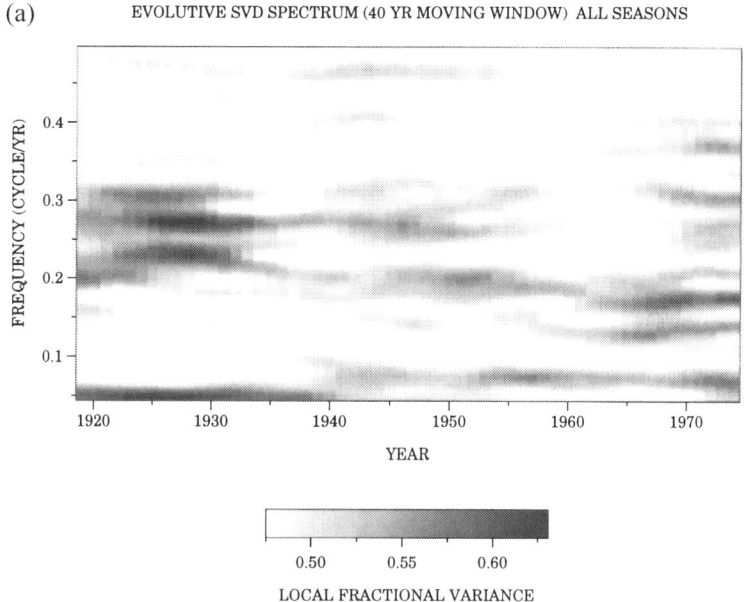

FIG. 29. Evolutive LFV spectrum based on performing the SVD analysis in a moving 40-year window for (a) all seasons, (b) cold-season, (c) warm-season. The amplitude of the spectrum as a function of time (center of moving window) and frequency is shown with the indicated grayscale, with confidence levels associated with the values given in Figure 28. Only the frequency range resolvable from secular variations ($f > 0.05$ for a 40-year moving window) is shown. [From Mann and Park (1996b).]

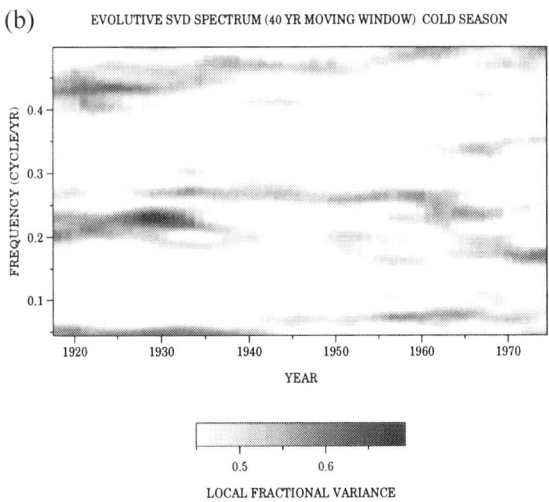

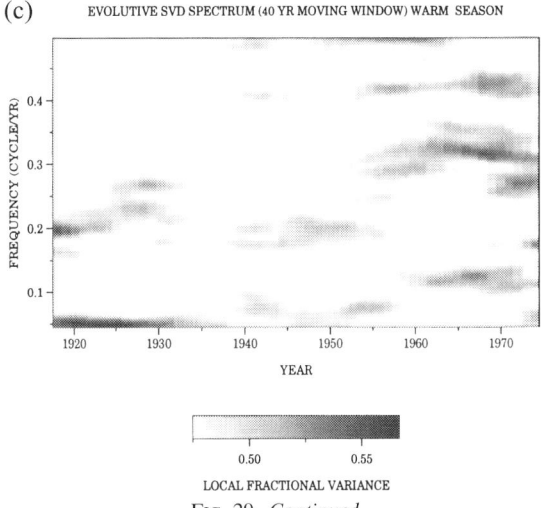

FIG. 29. *Continued.*

alternative wavelet generalization of the MTM-SVD approach (Lilly and Park, 1995; Park and Mann, 1998) allows for an automatic scaling of period and assumed phase-coherence time scale.

In the all-months analysis (Fig. 29a), there is clear evidence of organized frequency-domain structure within the ENSO band, with two dominant bands of variability clustered around 3- to 4-year-period and 5- to 7-year-period bands, that exhibit appreciable frequency and amplitude modulation. This behavior is *approximated* in the full-window analysis by amplitude-modulated statistically significant quasi-oscillatory components centered near 3-, 4-, 5-, and 7-year periods (see Section 4.1, ENSO Signal). A parsimonious picture of two distinct low-frequency (LF) and high-frequency (HF) bands of ENSO-related variance is thus indicated in the evolutive analysis. This description suggests a degree of frequency-domain organization which belies a simple *episodic* picture of ENSO, and reinforces the utility of a frequency-domain analysis of ENSO variability. The cold- and warm-season-only analyses show streaks of variance within these two same dominant bands, but the relative lack of frequency-domain structure in these cases relative to the all-months analysis implies a signal that is not simply phase-locked to the annual cycle. Nonetheless, the separate cold- and warm-season signal reconstructions are essential to understanding the relationship between tropical and extratropical expressions of the signal. The seasonal analyses substantiate the cold-season dominance of the quasi-biennial signal and reveal a drifting trend towards higher frequency during this century. This frequency modulation is consistent with the relationship that was observed between secular warming and the envelope of the *fixed-frequency* reconstruction of the quasi-biennial signal in the global temperature analysis (see Section 4.1, "Quasibiennial Signal" and Table IV).

Secular Signals

The primary secular mode (Fig. 30) accounts for 77% of the near zero-frequency variance in the joint SLP–temperature data set, and is associated with the secular trend in the global temperature data (i.e., the global warming signal of Section 4.1, Fig. 20). The warming pattern is evident in both cold and warm seasons, although certain regions show a marked seasonal dependence. For example, secular warming in eastern Asia and the North Atlantic is present only during the cold season, which might be related to winter land and sea-ice albedo effects.

FIG. 30. Spatial and temporal pattern of primary secular mode. (a) Time reconstructions for reference temperature gridpoints in the (i) central subtropical and (ii) eastern tropical Pacific, along with 2-year smoothed raw gridpoint data. (b) Cold-season and (c) warm-season spatial pattern. As in all similar subsequent spatial plots, temperature anomalies are indicated with the grayscale shown, with colour versions of the figures available electronically [ftp://eclogite.geo.umass.edu/pub/mann/ADVGEOPHYS-CLRFIGS/]. SLP patterns are contoured in units of millibars (mb), and reference gridpoints are indicated by a box (temperature gridpoint) or boxed "x" symbol (SLP gridpoint).

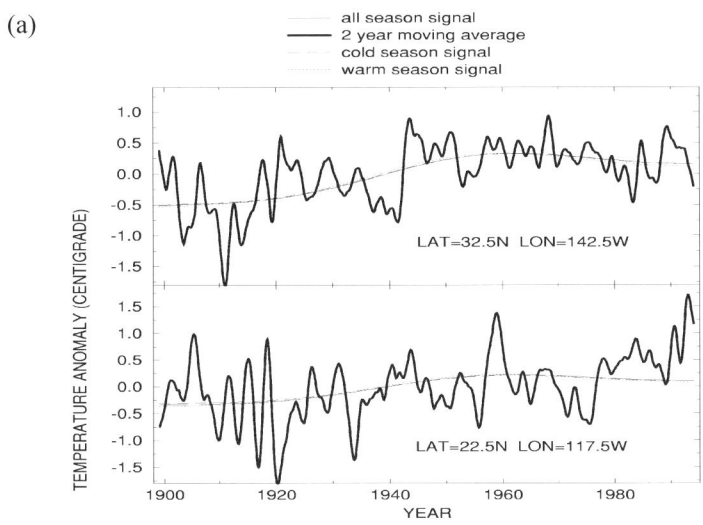

(a)

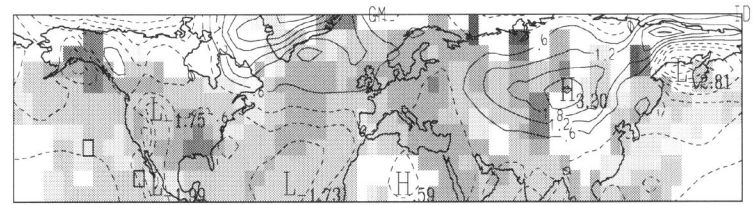

(b) COLD SEASON

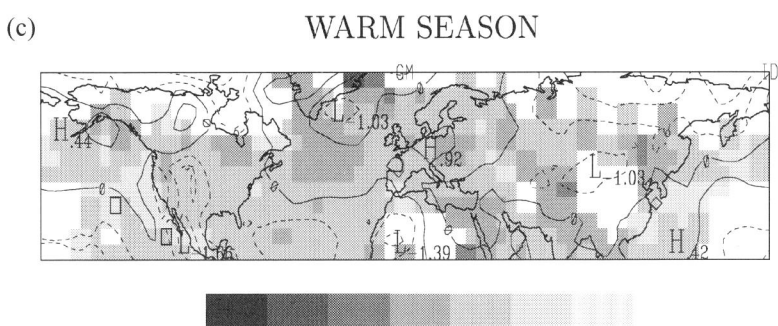

(c) WARM SEASON

The warming is coherent with regional SLP anomalies that imply altered atmospheric circulation patterns. In the southeastern U.S., the presence of enhanced cold-season cooling and warm-season warming might be related to anomalous circulation patterns (e.g., a trough-like trend over North America discussed elsewhere by Mann et al., 1995a) that favors a more continental influence in the region. Over the North Atlantic, a winter season north–south trend resembles the reverse of the NAO pattern (see Deser and Blackmon, 1993). This pattern could explain some of the asymmetry in warming along the western and eastern margins of the North Atlantic basin, although changes in heat transport by the Gulf Stream have also been speculated by Deser and Blackmon (1993). Qualitatively similar circulation anomalies have been variously observed in GCM simulations of the climatic response to greenhouse gas forcing (IPCC, 1996, chapter 6, 1996; Oglesby and Saltzman, 1992; Marshall et al., 1995). Other circulation anomalies are suggested. The questionable quality of the earlier SLP data, however, demands a cautious interpretation of long-term trends in SLP. The pronounced SLP anomalies near Northern Japan/Sea of Okhotsk, for example, correspond to the discontinuous early behavior of a small number of gridpoints.

The secondary secular mode (Fig. 31) accounts for a lesser, but nonetheless significant, 21% share of the secular band variance. This mode is associated with the multidecadal pattern of high-amplitude warming and subsequent cooling in the North Atlantic isolated in the global temperature data (Section 4.1). The objective time-domain signal-reconstruction procedure employed in this analysis favors a roughly 75-year-period "oscillation" time scale. This time scale, though poorly estimated in the context of a single secular variation, is consistent with persistent 50- to 100-year multidecadal/century-scale oscillations in proxy climate data (see Section 4.3).

Arguments for a relationship with the thermohaline circulation are further strengthened here by the joint relationship between sea surface temperatures and overlying SLP patterns. Delworth and collaborators (Delworth et al., 1993; 1997) have demonstrated that century-time-scale (40- to 80-year-period) oscillatory behavior can arise from climate mechanisms involving the thermohaline circulation, and perhaps coupled ocean–atmosphere processes, based on a 1000-year coupled ocean–atmosphere model simulation. In that study, a pattern of anomalously warm SSTs in the mid-latitude and polar North Atlantic, and weaker cold anomalies in parts of the tropical/subtropical North Atlantic, was associated with the enhanced-thermohaline circulation phase of the oscillation. The enhanced circulation state was in turn accompanied by a pattern of negative high-latitude and positive low-latitude SLP anomalies over the North Atlantic.

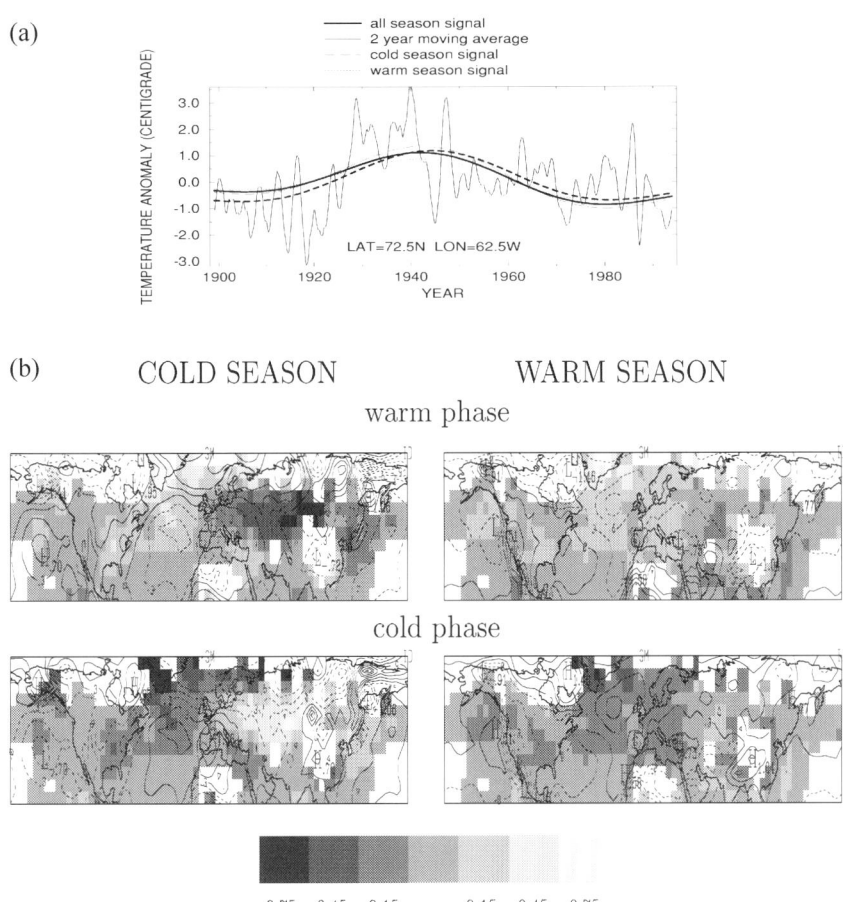

FIG. 31. Spatial and temporal pattern of secondary secular mode. (a) Time reconstructions for a reference temperature gridpoint in the North Atlantic, along with smoothed raw data series. An additional (thick dashed) curve shows the additional contribution of the primary secular mode to the total secular variation at this gridpoint. (b) Cold-season (left) and warm-season (right) spatial patterns of signal, showing in each case the phases of the signal corresponding to a warm (upper) and cold (lower) North Atlantic.

Whether the atmospheric pattern is simply a passive response or an intrinsic component to the underlying feedback system remains to be established. The opposite features were associated with the weakened state of thermohaline circulation.

We find additional support for such a coupled pattern in the multi-decadal signal isolated here. The North Atlantic regional features of the signal (consistent with those isolated in the associated global temperature pattern) confirm the comparison drawn by Kushnir (1994) between observed North Atlantic multidecadal variations and the multidecadal oscillations patterns of Delworth et al. (1993—henceforth DEL93). We find a pattern of surface temperature variability in the North Atlantic, evident in both cold and warm seasons, that resembles the surface temperature patterns of the DEL93 signal. Furthermore, consistent with DEL93, we find a persistent relationship between anomalous warm SSTs in the high-latitude North Atlantic and anomalous low SLP over part of the polar North Atlantic (Labrador Basin/Baffin Bay region). However, a *convincing* similarity is only found during the warm season, during which low pressure presides over the entire polar North Atlantic region during the warm SST phase of the signal. Thus, the agreement between the observed and modeled century-scale signal is imperfect. The substantial opposite-sign surface temperature anomalies over much of Eurasia during the cold season (note that the cold season dominates the pattern of the all-seasons global temperature analysis described by Fig. 20) appears to be related to a breakdown of mid-latitude westerlies in that region during the warm-North Atlantic phase inferred from the SLP pattern, associated with a decrease in their moderating influence on the cold-season Eurasian climate.

Interdecadal Signal

The interdecadal signal of Figure 22 is established as a significant signal in joint temperature and SLP over the Northern Hemisphere (Figs. 32–34), with a suggested Pacific basin center of activity. Although a number of possible mechanisms for such roughly bidecadal time-scale variability were discussed in Section 4.1, the joint-mode analysis provides evidence for the particular mechanism advocated in the modeling study of LB94. In that study, oscillatory $\sim$ 20-year-period behavior centered in the North Pacific was generated in a roughly 70-year simulation of a coupled ocean–atmosphere model. A delayed oscillator mechanism was argued for by LB94, involving feedbacks between gyre heat transport, changes in heat content and SST, and anomalous atmospheric circulation in the extratropical Pacific. The interdecadal signal isolated in the joint-field analysis exhibits remarkable resemblance in both temporal (i.e., both the $\sim$ 20-year period and the irregular modulated nature of the oscillation) and spatial (Fig. 33) characteristics to the LB94 signal. During the extreme of the cycle associated with anomalous negative oceanic heat content in the central North Pacific in LB94, for example, LB94 observe anomalous positive SSTs in the lower-latitude North Pacific. This anomalous SST pattern is associated with an increased latitudinal SST gradient, accompanying anomalous

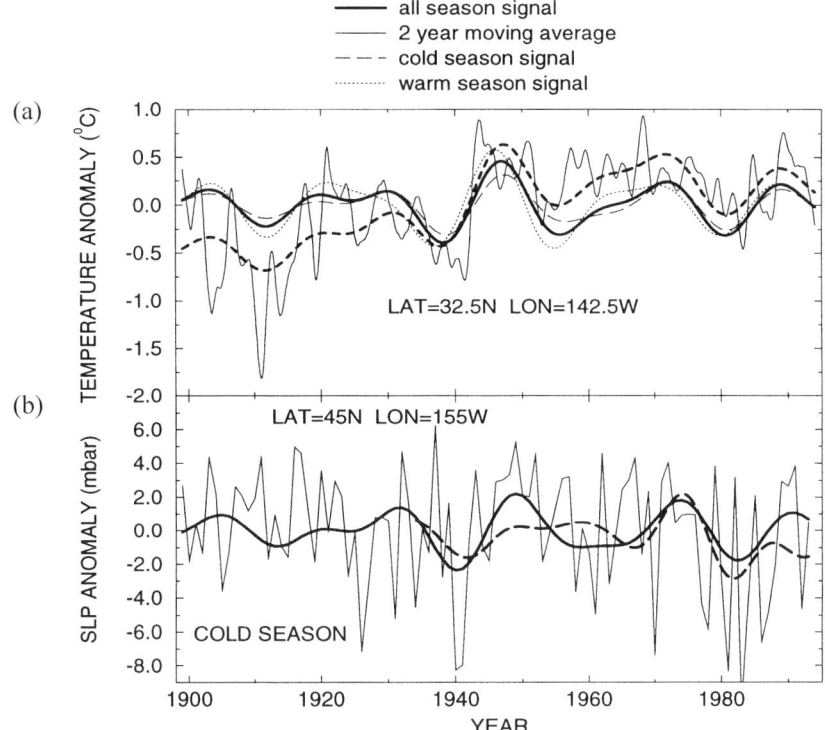

FIG. 32. Temporal pattern of interdecadal mode for (a) reference subtropical east/central Pacific temperature gridpoint and (b) reference mid-latitude central Pacific SLP gridpoint (cold season) along with the smoothed raw data series. The thick dashed line in (a) indicates the sum of the secular warming trend (see Fig. 24) and interdecadal mode. The thick long-dashed line in (b) shows the interdecadal and longer-term (lowpassed with a notch at 10-year period) variations in the SOI (scaled by a factor of 3 in mb) since continuous data is available, demonstrating an in-phase relationship of decadal-scale variations with the projection of the interdecadal signal onto cold-season winter North Pacific SLP. [From Mann and Park (1996b).]

low wintertime SLP over the North Pacific and inferred strengthening of mid-latitude westerlies. At the other extreme of the cycle, these patterns are reversed. This relationship is argued by LB94 to reflect a sequence of positive and negative feedbacks whereby changes in the meridional SST gradient force an equivalent barotropic response in the planetary wave structure. This altered circulation leads to changes in the latitudinal gradient in windstress curl, spinning down the subtropical North Pacific gyre, altering the poleward heat transport by the gyre and thus, the SSTs

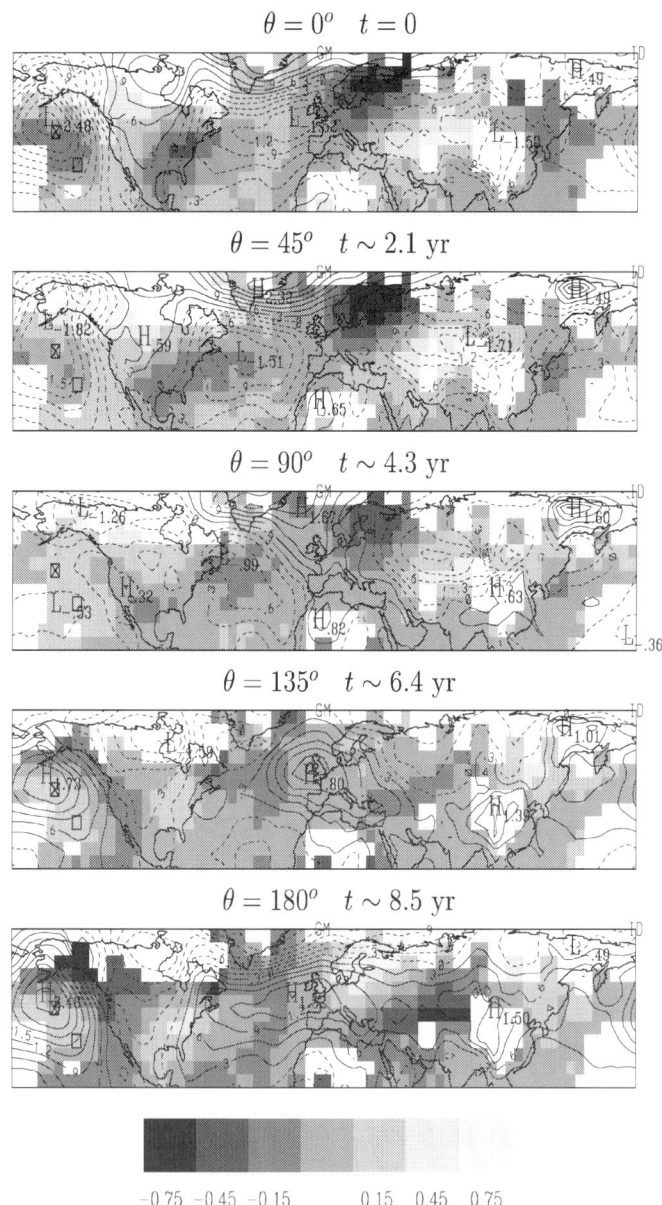

FIG. 33. Cold-season spatial pattern of interdecadal signal shown at progressive intervals, spanning one-half of a complete (~ 17-year) cycle. The absolute timing of relative phases of the pattern is defined by the reference temperature and slp series shown in Figure 32. [the signal progression here and in all similar following plots is opposite to that erroneously shown in Mann and Park (1996b).]

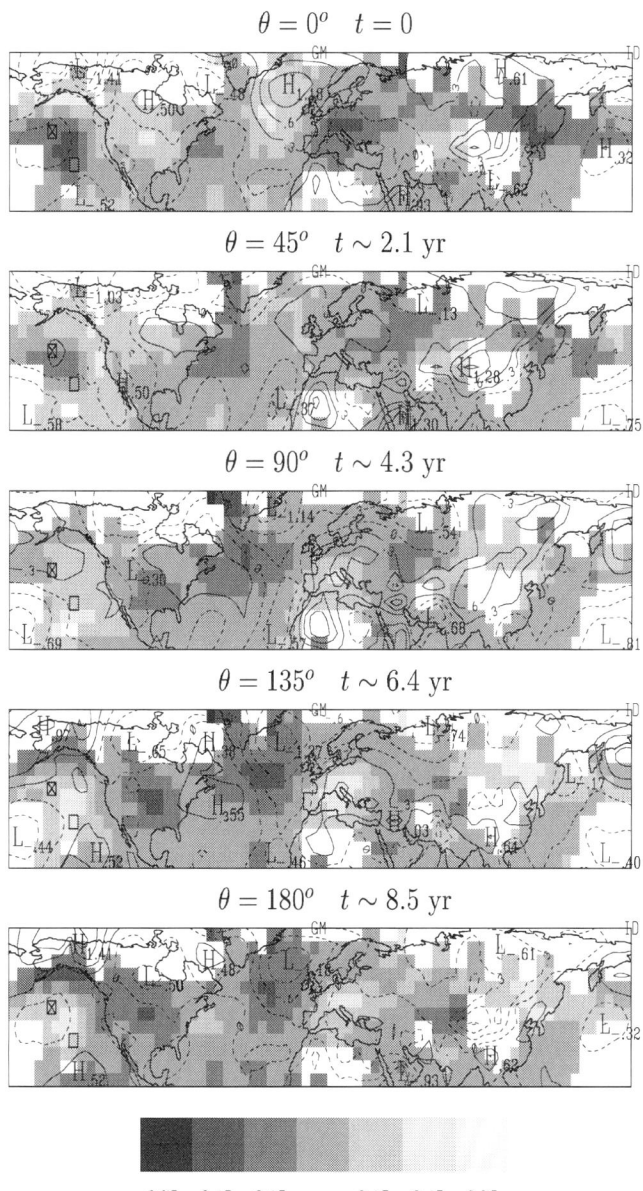

FIG. 34. Warm-season spatial pattern of interdecadal signal. Convention same as in Figure 33.

themselves. The mode is inferred to be an internal eigenmode that is excited by stochastic forcing.

In the cold season (Fig. 33), the peak positive (negative) SLP anomaly over the North Pacific is associated with a pattern over North America resembling the positive (negative) phase of the Pacific/North American teleconnection pattern (PNA—see Barnston and Livezey, 1986). The "PNA" signature of this interdecadal signal is distinct from the similar but spatially offset "TNH" pattern typically connected with the interannual ENSO signal (see "ENSO Signal" section below), consistent not only with LB94, but with other studies which have linked interdecadal variability with the PNA pattern (e.g., Trenberth, 1990; Mann and Park, 1993). The ~ 20-year time scale of an oscillatory cycle is attributed by LB94 to the intrinsic spinup time scale of the North Pacific gyre, with the adjustment in gyre heat transport lagging windstress changes by several years, thus providing feedbacks that support oscillatory behavior. The warm-season expression of the signal (Fig. 34) emphasizes the seasonal persistence (and consistency) of SST anomalies in the Pacific which appear to lead to cold-season-specific excitation of high-amplitude extratropical circulation anomalies. This seasonal distinction is also consistent with the LB94 mechanism. One notable exception is an implied strengthening and weakening of the Asian monsoonal pattern associated with seasonally opposite SLP variations over central Asia.

While some compelling similarities are found with the simulations of LB94, there are notable discrepancies as well. For example, the interdecadal pattern of LB94 is strongest over the western North Pacific where the model SST gradient is largest. In contrast, our observed pattern exhibits the largest SST gradient anomalies in the eastern North Pacific. Furthermore, high-amplitude variability in the Atlantic region and elsewhere suggests strong downstream teleconnections or perhaps even coupling with Atlantic basin processes. Such issues are discussed further for the "Quasi-decadal Signal" below. This monsoonal pattern discussed above is just one of several apparent connections with low-frequency variability in ENSO which are also not explained by the extratropical mechanism of LB94, and may indicate the added importance of coupling between the extratropics and the tropics. This relationship is evident not only in the ENSO-like patterns of warming and cooling (see also Section 4.1) but in the mild positive east–west SLP gradient across the tropical Pacific coinciding with anomalous warmth in the eastern tropical/subtropical Pacific. Furthermore, a close association between the signal and decadal-scale variations in the southern oscillation (SO—see Fig. 32 and also Trenberth, 1990) and NINO3 SST indices (both exhibit spectral coherences significant at ~ 95% confidence level within the interdecadal frequency band) is evident. Such connections suggest some relationship with low-frequency changes in ENSO. It is possible that the high-amplitude extratropical variations may force a weaker modulation of the tropical Pacific, thus impacting the ENSO phenomenon. Conversely, a more complicated coupled tropical/extratropical mechanism could be at work, perhaps combining dynamical mechanisms explored by both LB94 and Gu and Philander (1997).

In terrestrial regions, the temperature patterns appear consistent with the effects of altered atmospheric circulation on sensible heat redistribution. For example, the cold anomaly in the southeastern U.S. and the warm anomaly in the northwestern U.S. are consistent with sensible heat transport by the anomalous PNA-type pattern. This phase of the cycle is associated with a significant share of the general hemispheric (and in fact global, as shown in Section 4.1) warmth that was observed during the mid-to-late 1980s (as well as the late 1940s and early 1970s), but the signal projects a tendency for negative temperature anomalies during the early-to-mid 1990s, counter to the continued acceleration of warming which, as discussed earlier, may relate to nonstationary behavior in ENSO (Trenberth and Hoar, 1995).

Quasi-Decadal Signal

In contrast to the interdecadal signal discussed above, the quasi-decadal (10- to 11-year-period) signal (Figs. 35 and 36), as its counterpart in global temperature data (Fig. 23), appears to be tied more closely to the North Atlantic region. Weaker variability is observed throughout the remaining Northern Hemisphere. The more widespread global temperature data of Section 4.1 supports a more distinct separation (i.e., a "spectral gap"— compare Figs. 18 and 28) between the quasi-decadal and interdecadal signals than does the joint-field analysis. With the short duration of instrumental data available and only a Northern Hemisphere domain, it is more difficult to distinguish adequately between the statistical models of independent interdecadal and quasi-decadal spatially distinct processes, and a single more broadband, coupled basin signal.

The quasi-decadal signal at certain phases exhibits an NAO-like SLP pattern (Fig. 36) consistent with observations of quasi-decadal variability in the NAO index itself (Hurrell, 1995). However, the SLP pattern is more monopole, with a higher-amplitude center of variation in the subpolar North Atlantic than in the subtropics. This distinction underscores the danger of analyzing simple indices such as the NAO for inferences into low-frequency climatic variability; while the pattern of Figure 36 projects onto an NAO or other diagnostic index of North−South SLP gradients in

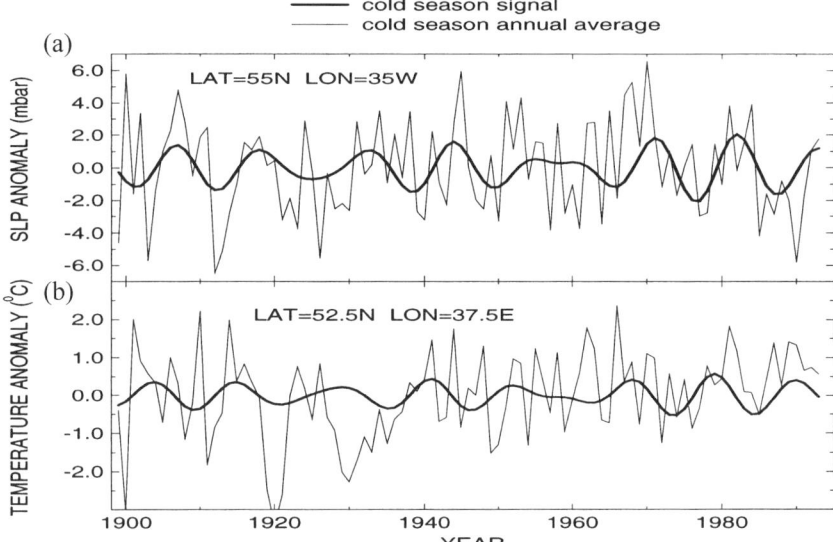

FIG. 35. Time reconstruction of ~ 10- to 11-year quasi-decadal signal for (a) reference mid-latitude/subpolar North Atlantic SLP gridpoint (cold season) and (b) similar latitude western Soviet temperature gridpoint. Note the 90° phase lag between the two variations, as expected from the discussion in the text. [From Mann and Park (1996b).]

the North Atlantic, it does not in fact resemble the classic NAO pattern (e.g., Rogers, 1984; Lamb and Peppler, 1987). The observed pattern, furthermore, is not associated with a standing SLP dipole, but rather with a time-evolving SLP anomaly pattern. Cold conditions over Northern Europe/Western Eurasia, for example, are consistent with the inferred breakdown of maritime influence due to anomalous low SLP centered over Great Britain during the third snapshot shown. This pattern is almost orthogonal to an NAO anomaly pattern.

Both temperature and SLP signatures over the North Atlantic are similar to that identified by Deser and Blackmon (1993) for quasi-decadal North Atlantic climate variations, but the signal appears to exhibit a wider hemisphere-scale influence. Mann *et al.* (1995a) noted that the quasi-decadal signal dominates the low-frequency winter circulation variability in the Great Basin (compare Fig. 36 in this region). They were able to demonstrate an association between the anomalous low pressure over the region during the phase of the third snapshot of Figure 36 and coincident decreased storm activity/increased regional precipitation anomalies leading to an increasing trend in the Great Salt Lake volume. In contrast with

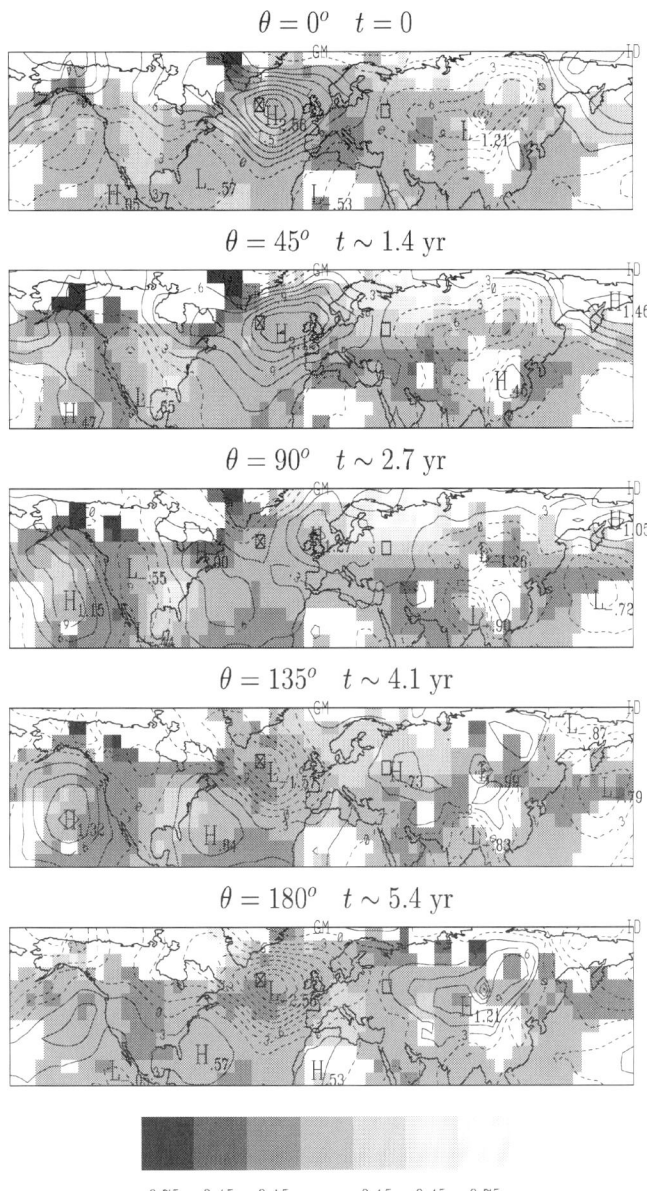

FIG. 36. Cold-season spatial pattern of quasi-decadal signal spanning one-half of complete (~ 11-year) cycle. The initial snapshot corresponds to the peak positive winter SLP anomaly in the North Atlantic.

the PNA circulation anomalies associated with the interdecadal signal discussed above, which are interpreted as a downstream perturbation set up by Pacific basin climate variations, we infer here an origin in the high-amplitude variations centered in the Atlantic. The notion that an Atlantic source of variability would have such strong upstream impacts is counterintuitive, but nonetheless consistent with studies that have established the importance of retrogressing long-wave disturbances during the cold season over North America (e.g., Lanzante, 1990).

No connection is found between the quasi-decadal signal and ENSO. Spectral coherences with quasi-decadal band variations in NINO3 and the SOI are statistically insignificant. The purely cold-season nature of the signal in this analysis seems to arise from the fact that its most prominent features—i.e., the extratropical circulation and surface temperature anomalies in the North Atlantic—are cold-season dominant. The signal may persist year-round in the tropics, which are only weakly represented by our largely extratropical spatial sampling. The cold-season expression in the extratropics may also indicate a carrier signal beneath the seasonal thermocline, only expressed at the surface during the winter period when stratification is weak and deep convection occurs (see, e.g., Dickson, 1997 and references therein).

LB94 suggest that a mechanism similar to that discussed for the interdecadal Pacific-centered signal, combined with a narrow basin geometry, could lead to an analogous shorter-period, quasi-decadal signal centered in the Atlantic. Our study offers some limited support for this hypothesis, as the relationship between the evolving patterns of SLP and SST in the North Atlantic shows some of the same features. For example, the initial snapshot of Figure 36 exhibits a consistent pattern of anomalous high pressure associated with inferred weakened westerlies in mid-latitudes, and a decreased SST gradient over much of the Atlantic. However, the mechanism of LB94 does not explain some very important features of the signal, most notably the tropical SST dipole associated with the larger-scale expression of the signal as shown earlier in Figure 23.

A number of mechanisms besides that of LB94 could explain such features or even the signal itself. Some researchers indicate the importance of low-frequency advection of salinity anomalies by gyres (Weaver *et al.*, 1991) or the interaction of thermal and salinity anomalies (Yang and Neelin, 1993) on deep water production and tropical ocean–atmosphere interactions (Mehta and Delworth, 1995). Chang *et al.* (1997) present a simple, but compelling, thermodynamic coupled ocean–atmosphere mechanism which leads to a quasi-decadal oscillating SST dipole (and corresponding wind anomalies) in the tropical Atlantic. Tourre *et al.* (1998) propose how the interaction of such tropical anomalies with other processes, such as gyre-scale advection, can lead to a full-basin expression of signals that originate in the tropical Atlantic.

Several studies have also presented possible connections between the ~ 11-year solar cycle and quasi-decadal climate fluctuations. Such statistical relationships have typically been established with stratospheric and upper tropospheric fluctuations (e.g., Labitzke and van Loon, 1988; Tinsley, 1988), rather than with lower atmospheric or surface climate data. While we do not find evidence for a statistically significant correlation with the ~ 11-year sunspot cycle in this quasi-decadal *surface* climate signal, there is some evidence for phase-locking during the past few decades (see also Mann *et al.*, 1995b) coinciding with a period of high-amplitude solar cycle variations. The argument that an internal quasi-decadal eigenmode could resonate with such external forcing, given sufficient amplitude, cannot be ruled out. However, the direct forcing (i.e., the irradiance changes) associated with such solar variations is small (see, e.g., Lean *et al.*, 1995) and some as yet unestablished means of amplification of the forcing (e.g., cloud electrification—Tinsley, 1988, or modulation of stratospheric ozone concentrations—Robock, 1996) would be required for a viable mechanism.

ENSO Signal

Although the region sampled in this analysis does not include the equator or Southern Hemisphere, enough of the tropical Northern Hemisphere is included that relationships with ENSO can be examined. The domain includes a sizable region known to experience the direct effects of both El Niño and the Southern Oscillation or "SO" (see, e.g., Trenberth and Shea, 1987). Our analysis thus complements similar previous analyses of ENSO-related climate variability (Barnett, 1991) where tropical ENSO-band climate patterns were examined, by using a long, widespread, though less tropics-dominated data set.

The time-domain signal is reconstructed based on the evolutive procedure (employing a 40-year moving window) recognizing the distinct low-frequency "'LF" and high-frequency "HF" band components isolated in the evolutive spectrum (see Fig. 29). The reference signal is shown for a temperature gridpoint within the reach of the warm "El Niño"/cold "La Niña" tongue, and an SLP gridpoint within the limits of the Indonesian convective region strongly impacted by the Southern Oscillation (Fig. 37). These two reference series are analogous to NINO3 and reverse SOI indices, and as expected, are roughly in phase. Both HF and LF time-domain components exhibit highly significant spectral coherence within their respective frequency bands (95–99% confidence level) with both the

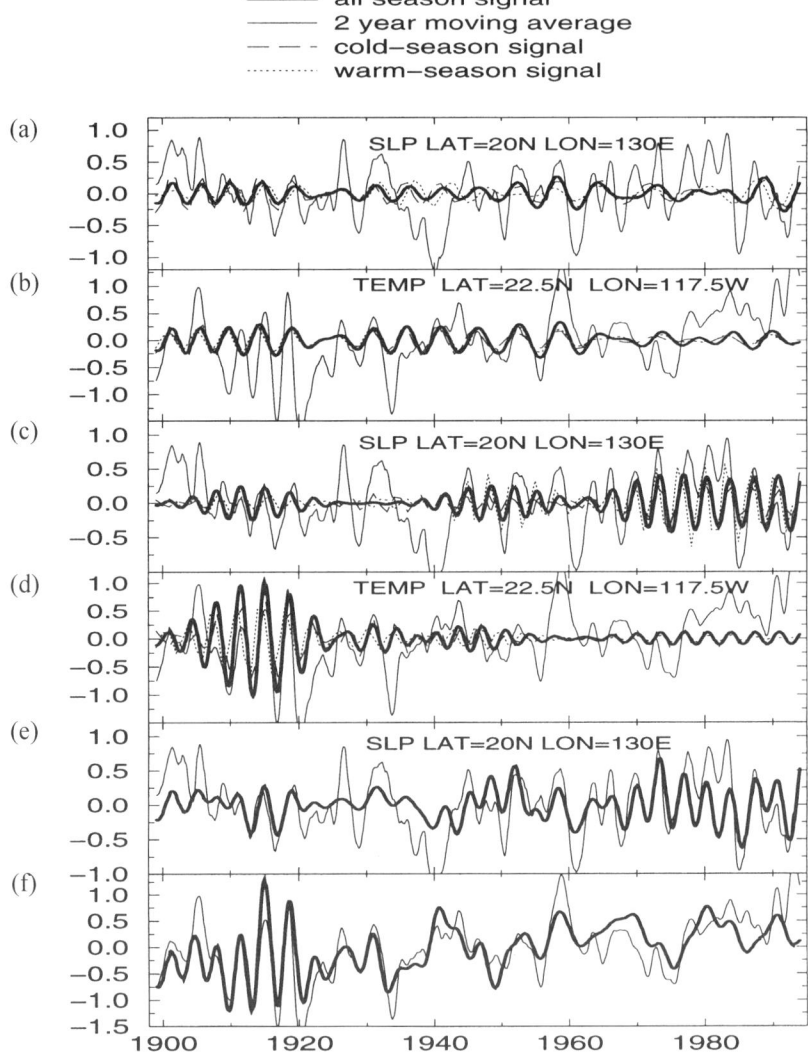

FIG. 37. Time reconstruction for (a) low-frequency component, reference SLP gridpoint in the tropical western Pacific, (b) low-frequency component, reference temperature gridpoint in the eastern tropical Pacific, (c) high-frequency component, same SLP gridpoint, (d) high-frequency, same temperature gridpoint, (e) sum of interdecadal and both ENSO components, SLP gridpoint, (f) sum of interdecadal and both components + secular warming signal, temperature gridpoint. [From Mann and Park (1996b).]

standard SOI and NINO3 indices, providing independent evidence of a direct relationship with the ENSO signal. The spatial patterns for the LF component are shown for both cold (Fig. 38) and warm (Fig. 39) seasons. For simplicity, we show the spatial evolution for only one (the "low-frequency") of the two dominant ENSO-related signals, and for a "composite" cycle obtained as the average pattern of evolution of the signal over the 95-year period. While certain features differ between the two signal components, we focus below only on those features that are shared by the low- and high-frequency patterns and persist from cycle to cycle over the slowly evolving patterns of reconstruction. Low-frequency and secular changes in the regional pattern of the large-scale ENSO signal may nonetheless exist (e.g., Ropelewski and Halpert, 1987) and should not be dismissed as stochastic event-to-event variations. Consistent with such nonstationarity in the patterns of ENSO, we note that the relative signature of "El Niño" vs "SO" characteristics is somewhat variable over time for the higher-frequency ENSO component, as evident in the high-amplitude temperature fluctuations in the tropical eastern Pacific early in the century contrasting with relatively high-amplitude SLP fluctuations in the western tropical Pacific later (Fig. 37). The nearly constant phase relationship (i.e., zero lag) between these SLP and temperature variations indicates a consistent "ENSO" signal. Studies of long-term proxy data suggest similar nonstationarity in the characteristics of ENSO over several centuries (Cole *et al.*, 1993; Linsley *et al.*, 1994; Dunbar *et al.*, 1994; Bradley *et al.*, 1994).

The initial stage shown for both the cold- and the warm-season signal coincides with the peak low phase of the Southern Oscillation (maximum SLP anomaly in the Indonesian convective region, minimum SLP anomaly in the eastern Pacific) in phase with El Niño conditions (maximum positive SST anomalies along the eastern tropical and subtropical Pacific coast). Consistent with other studies (e.g., Horel and Wallace, 1981), the low phase of the Southern Oscillation is observed to accompany notable cold-season circulation anomalies over the North Pacific and North America. The cold-season pattern over North America during this positive ENSO phase—e.g., roughly the first two stages shown—most closely resembles the Tropical/Northern Hemisphere (TNH) pattern which recent studies have shown to be a characteristic cold-season extratropical teleconnection of ENSO (Livezey and Mo, 1987), with patterns resembling the NAO (e.g., Rogers, 1984) and the Western Pacific Oscillation (WPO) somewhat evident. There is a predominant tendency for hemisphere-wide warmth at this stage, although cooling is found in certain areas where cold advection is suggested or, in the case of Greenland, where cooling under a cold-season high-pressure region is suggestive of enhanced radiational

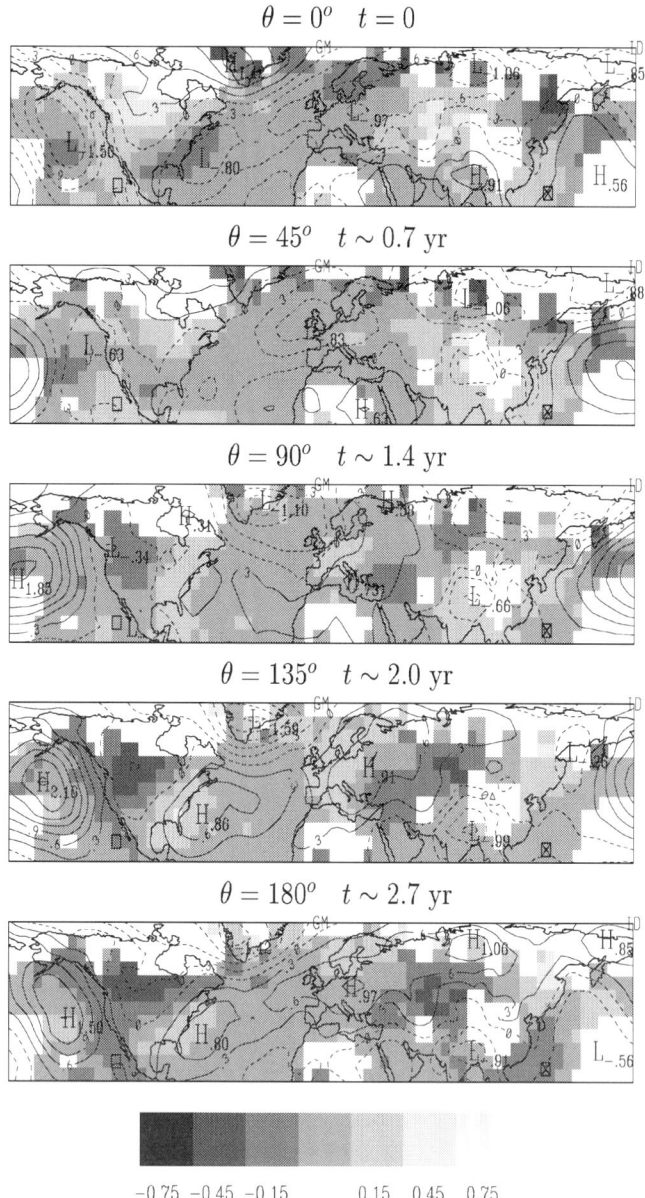

FIG. 38. Canonical cold-season spatial pattern of low-frequency ENSO signal spanning one-half of the average ~ 5.4-year cycle length. The initial snapshot corresponds to peak or near-peak (El Niño/low SO) ENSO conditions in the cycle.

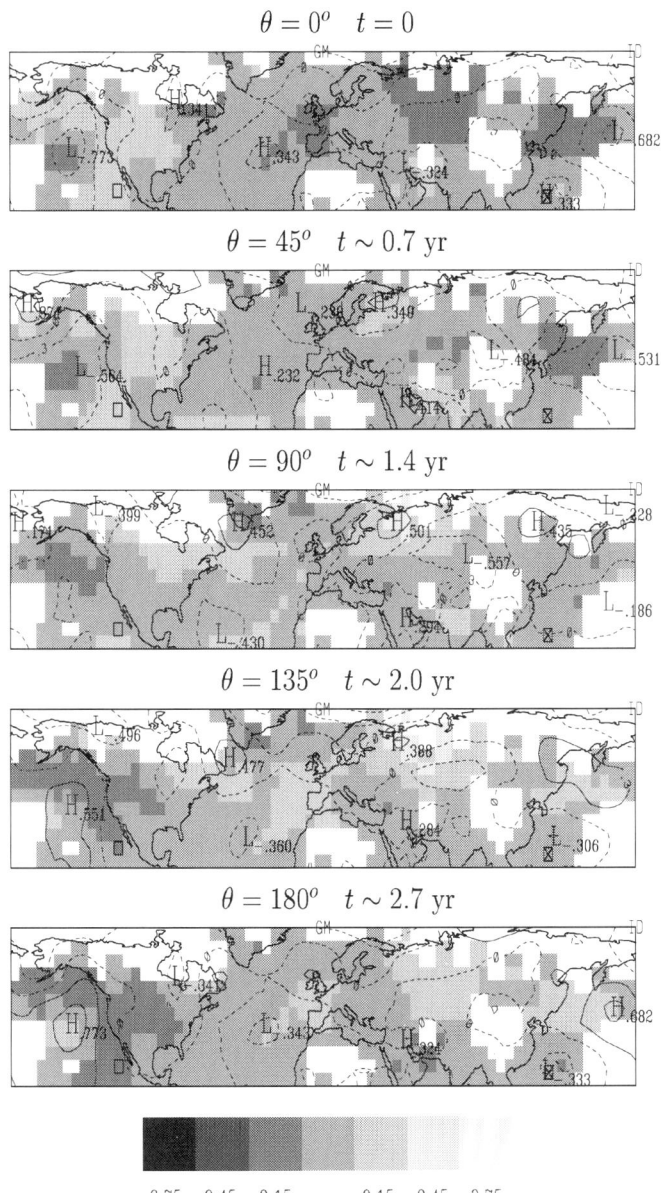

FIG. 39. Canonical warm-season spatial pattern of low-frequency ENSO signal.

cooling. As the cycle progresses, by the third stage shown (halfway between low/high SO and El Niño/La Niña conditions), the TNH pattern over North America has broken down (although a considerable low-pressure anomaly remains over the North Pacific) and temperature anomalies are generally weak. The cycle subsequently progresses to the reverse of the initial phase, associated with high SO, La Niña, a reverse TNH pattern, and predominant coolness over the hemisphere.

The SST anomaly patterns are consistent between, and quite persistent through the warm season, but tropical and extratropical circulation anomalies (including the east–west SLP gradients in the tropical Pacific) are considerably weaker. Accordingly, land surface temperature anomalies typically are lower amplitude. In places (e.g., central Asia), temperature anomalies in the distinct seasons are of opposite sign due to the seasonal specificity of inferred circulation anomalies. The seasonal persistence of the signal thus appears to arise largely from the year-round persistence of tropical SST anomalies during the evolution of the signal, while atmospheric circulation anomalies are more seasonally variable.

Quasi-Biennial Signal

The joint-mode analysis confirms the association of the large-scale surface quasi-biennial signal (Figs. 40 and 41) with a distinct NAO SLP pattern as in Trenberth and Shin (1984). Hemisphere-wide teleconnections are nonetheless evident. The time-domain signal (Fig. 40—performed using the evolutive reconstruction procedure with a 20-year window) exhibits considerable amplitude modulation on decadal time scales, which may indicate a coupling with other lower-frequency variability discussed earlier. This signal is also observed to be highly phase-coherent with no phase modulation, which, as its period is not an integral multiple of the annual cycle, indicates weak if any phase-locking to the annual cycle. Nonetheless, the signal features are most distinct during the cold season (Fig. 41). The phase shown in the initial snapshot, associated with the positive NAO pattern over the Atlantic, also shows a "lopsided" dipole SLP pattern over the Pacific. The temperature pattern follows expected patterns of sensible heat transport with warm (cold) anomalies associated with regions of implied southerly (northerly) advection and, over extratropical Eurasia, enhanced warmth due to strengthened westerlies and an associated moderated winter climate. Similar relationships between circulation and surface temperature anomalies are found as the signal evolves over a typical cycle. The lack of large temperature anomalies in the eastern Pacific or any sizable east–west SLP gradients in the tropical Pacific during the evolution of the signal would seem to cast doubt on a

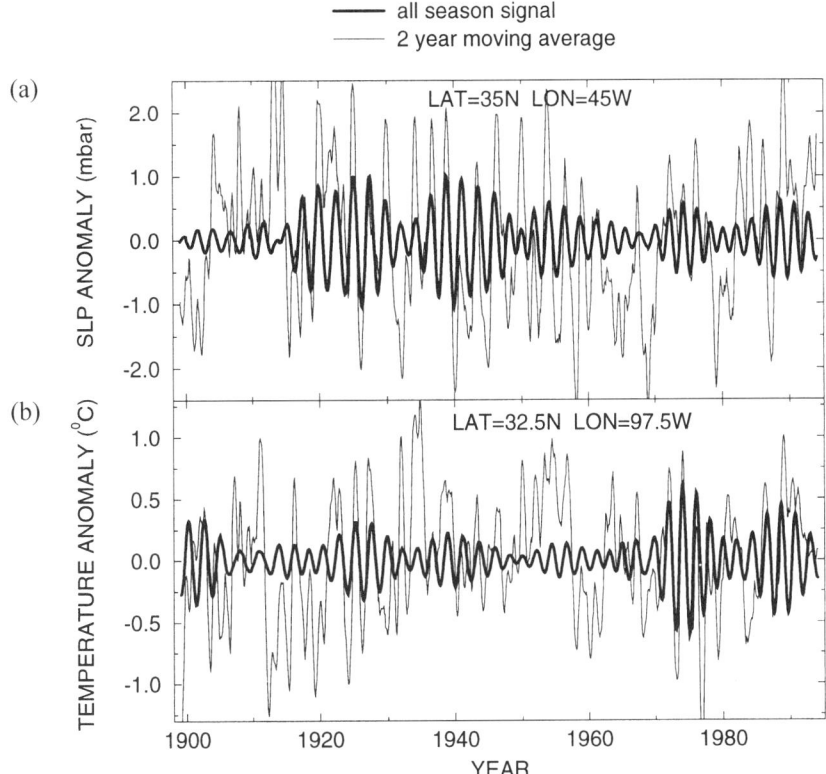

FIG. 40. Time reconstruction of ~ 2.1- to 2.2-year quasi-biennial signal (all months) for (a) reference SLP gridpoint in central subtropical North Atlantic and (b) reference temperature gridpoint in western Soviet Union. [From Mann and Park (1996b).]

direct connection with ENSO. However, we *do*, somewhat paradoxically, find significant spectral coherences between the quasi-biennial signal and quasi-biennial-band (QB) fluctuations in ENSO (that is, both the SOI and NINO3—see also Trenberth and Shin, 1984; Barnston *et al.*, 1991). This relationship could thus reflect a quasi-biennial extratropical forcing of weaker fluctuations in ENSO. This inference is not dissimilar from that of Barnett (1991), who noted that QB-band variations, although apparently somewhat related to ENSO, had a far weaker loading in the tropical Pacific than lower-frequency ENSO variability. Consistent with Barnett (1991) and Trenberth and Shin (1984), no significant relationship is observed with the quasi-biennial oscillation in equatorial stratospheric wind data available for the past few decades (see Naujokat, 1986), with spectral

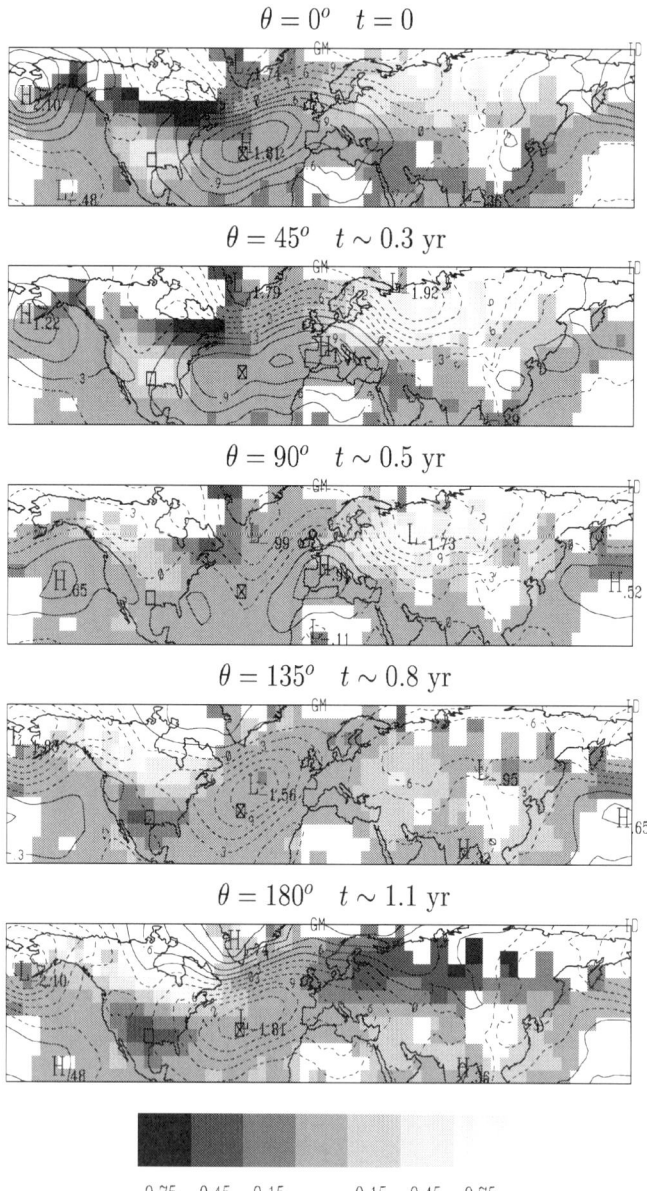

FIG. 41. Canonical cold-season spatial pattern of quasi-biennial signal, spanning one-half of the average ~ 2.2-year cycle length. The initial snapshot corresponds to the peak positive NAO pattern over the North Atlantic.

coherences in the quasi-biennial band that barely breach the median confidence level for significance.

Summary

MTM-SVD analysis of a joint twentieth-century temperature-anomaly/ sea-level pressure data set, restricted to the Northern Hemisphere, lends strong support to the existence of the oscillatory climate signals, identified in the temperature-anomaly data set alone. The interdecadal 16- to 18-year climate signal appears consistent with a gyre spinup and mid-latitude ocean–atmosphere interaction, a mechanism predicted in a recent coupled ocean–atmosphere simulation. Weaker quasi-decadal (10- to 11-year time scale), largely cold-season oscillatory behavior is more closely tied to the North Atlantic and may involve analogous mechanisms. Rather than modulating a simple NAO pattern, the decadal oscillation is associated as well with a lateral migration of pressure and temperature anomalies, behavior that would be missed in a simple time series of a fixed NAO index. Interannual variability is examined with an "evolutive" generalization of our procedure to capture the time-evolving frequency and amplitude characteristics of the associated climate signal. Variability exhibiting the characteristic climatic patterns of the global El Niño/Southern Oscillation (ENSO) phenomenon is described by two largely distinct frequency bands within the broader 3- to 7-year ENSO band. The drifting central frequencies of these two dominant bands are suggestive of nonstationary behavior in ENSO. A quasi-biennial signal exhibits a gradual trend towards increasing frequency.

4.3. Long-Term Multiproxy Temperature Data

To more confidently identify spatiotemporal climate signals at interdecadal and longer time scales, we make use of a small (35), but globally distributed, set of high-resolution temperature proxy reconstructions (Fig. 42) available for most of the past half-millennium. These data include tropical (e.g., Thompson, 1992) and extratropical ice melt measurements (Bradley and Jones, 1993), tropical corals (Cole *et al.*, 1993; (Dunbar *et al.*, 1994), dendroclimatic reconstructions (Jacoby and D'arrigo, 1989; Bradley and Jones, 1993), and a handful of very long historical sources (Bradley and Jones, 1993). Although extratropical records primarily reflect warm-season climatic variations (see Bradley and Jones, 1993), the signals of interest (see, e.g., Section 4.2, Fig. 28) are believed to be seasonally robust. The characteristics of the different proxy records employed in the network are described in Table V.

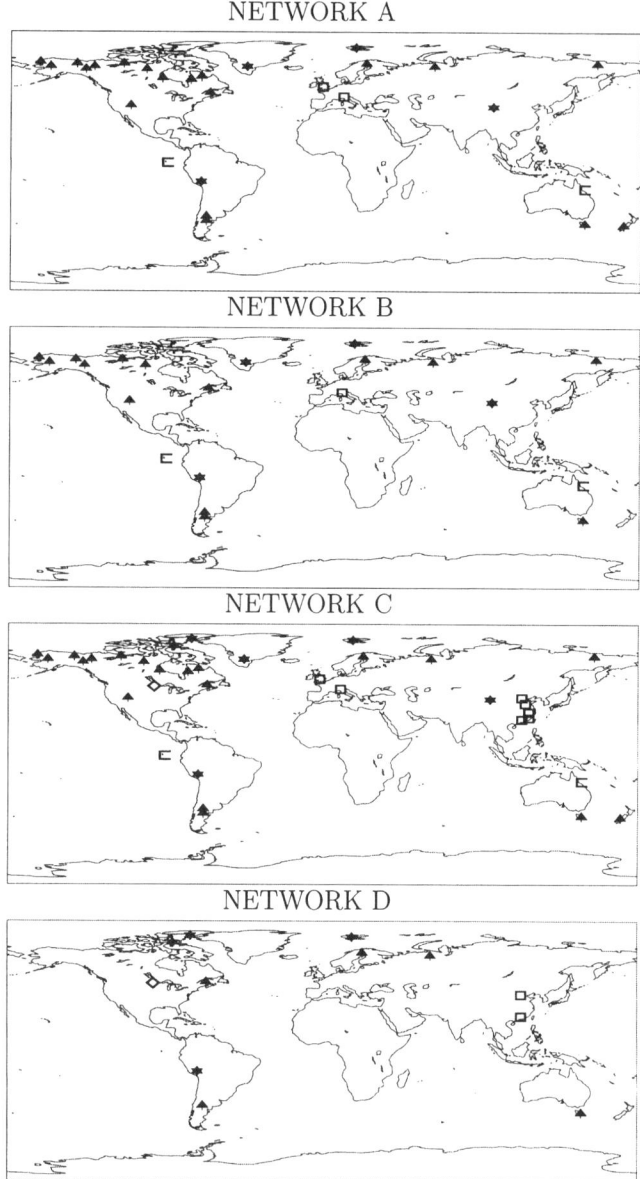

FIG. 42. Distribution of proxy temperature reconstructions used in the present study for experiments (a), (b), (c), and (d). Squares denote historical or instrumental records, umbrella or "tree" symbols denote dendroclimatic reconstructions, "⊏" symbols indicate corals, and diamond indicates varved lake sediment record. [Reprinted with permission from Mann *et al* (1995b). *Nature* (London), 378, 266–270. Copyright ©1995 Macmillan Journals Limited.]

TABLE V Proxy Data Records Used: Description / Type of Record, Location
(Latitude, Longitude), Beginning and Ending Year, Time Resolution, and Reference

Record	Location	Begin	End	Resolution	Reference
1. Northern treeline (tree ring)	69N, 197E	1515	1982	annual	Jacoby and D'Arrigo
2. Northern treeline (tree ring)	66N, 203E	1515	1982	annual	Jacoby and D'Arrigo
3. Northern treeline (tree ring)	68N, 218E	1515	1982	annual	Jacoby and D'Arrigo
4. Northern treeline (tree ring)	64N, 223E	1515	1982	annual	Jacoby and D'Arrigo
5. Northern treeline (tree ring)	66N, 228E	1515	1982	annual	Jacoby and D'Arrigo
6. Northern treeline (tree ring)	68N, 245E	1515	1982	annual	Jacoby and D'Arrigo
7. Northern treeline (tree ring)	64N, 258E	1515	1982	annual	Jacoby and D'Arrigo
8. Northern treeline (tree ring)	58N, 267E	1515	1982	annual	Jacoby and D'Arrigo
9. Northern treeline (tree ring)	57N, 283E	1515	1982	annual	Jacoby and D'Arrigo
10. Northern treeline (tree ring)	59N, 289E	1515	1982	annual	Jacoby and D'Arrigo
11. Northern treeline (tree ring)	48N, 294E	1515	1982	annual	Jacoby and D'Arrigo
12. Western U.S. (tree ring)	42N, 249E	1600	1982	annual	Bradley and Jones
13. Northern Patagonia (tree ring)	38S, 292E	869	1983	annual	Bradley and Jones
14. Central Patagonia (tree ring)	41S, 292E	1500	1974	annual	Bradley and Jones
15. N. Scandinavia (tree ring)	68N, 23E	500	1980	annual	Bradley and Jones
16. Northern Urals (tree ring)	66N, 62E	1400	1969	annual	Bradley and Jones
17. Upper Kolyma River (tree ring)	68N, 155E	1550	1977	annual	(Earle, personnal communication)
18. Tasmania (tree ring)	43S, 148E	900	1989	annual	Bradley and Jones
19. South New Zealand (tree ring)	44S, 170E	1730	1978	annual	Bradley and Jones
20. Agassiz (ice melt)	81N, 280E	466	1966	5 yr	Bradley and Jones
21. Southern Greenland (ice melt)	66N, 315E	1545	1988	annual	Bradley and Jones
22. Devon (ice melt)	75N, 275E	1400	1970	5 yr	Bradley and Jones
23. Svalbard (ice melt)	79N, 17E	1400	1985	annual	Bradley and Jones
24. Quelccaya (ice core, O^{18})	14S, 289E	470	1984	annual	Thompson
25. Dunde (ice core, O^{18})	38N, 96E	1606	1985	annual	Thompson
26. Central England (instrumental)	52N, 358E	1730	1987	annual	Bradley and Jones
27. Central Europe (historical)	45N, 10E	1550	1979	annual	Bradley and Jones
28. Eastern China (historical)	24N, 114E	1380	1980	decadal	Bradley and Jones
29. Northern China (historical)	39N, 118E	1380	1980	decadal	Bradley and Jones
30. Yellow River (historical)	35N, 116E	1470	1980	decadal	Bradley and Jones
31. S.E. China (historical)	25N, 118E	1470	1980	decadal	Bradley and Jones
32. Yangtze River (historical)	29N, 118E	1470	1980	decadal	Bradley and Jones
33. Galapagos (coral, O^{18})	1S, 270E	1607	1981	annual	Dunbar et al.
34. Great Barrier Reef (coral)	19S, 148E	1615	1982	annual	Lough (personal communication)
35. Minnesota (lake varve)	48N, 259E	980	1960	decadal	Bradley and Jones

To trade off the limitations of temporal resolution and duration (see Table V) with that of spatial coverage, it was useful to analyze four subsets of the data independently: (A) 27 shorter records (1730–1969) with annual resolution, (B) 21 medium-duration (1615–1969) records with annual resolution, (C) 35 shorter (1730–1960) records with decadal resolution, and (D) 12 longer (1400–1960) records with decadal resolution.

LFV Spectra

First, we compared the proxy network against instrumental data in sampling and detecting the same large-scale temperature signals. We performed parallel analyses of data during the last century based on (i) the global distribution of instrumental temperature data from 1890–1989 analyzed in Section 4.1, (ii) proxy group A over the abbreviated interval 1890–1969, and (iii) a sparse subsampling of the instrumental data of (i) chosen to mimic the spatial distribution and seasonal sampling and 1890–1969 time interval of the proxy data set A. A paucity of high-latitude instrumental sites impedes a perfect spatial match between data sets (ii) and (iii). The associated LFV spectra are shown in Figure 43. All four data sets exhibit a statistically significant 16- to 18-year-time-scale interdecadal signal ($f \approx 0.06$ cycle/yr) significant at the 95% confidence level. The quasi-decadal signal of Sections 4.1 and 4.2 ($f \approx 0.09$ cycle/yr) is at least marginally significant ($> 90\%$ level) in each of the data sets. Though the quasi-decadal signal is cold-season dominant in the extratropical Northern Hemisphere (Section 4.2, "Quasidecadal Signal"), the global sampling of the proxy network probably facilitates its detection with predominantly warm-season proxy indicators. Within the secular regime, both the primary and the secondary mode are significant, as in the observational temperature and joint temperature/SLP analyses of Sections 4.1 and 4.2, respectively. This exercise indicates that the proxy data network appears to be capable of capturing large-scale climatic processes evident in recent instrumental-based analyses, though with inexact calibration.

Figure 44 shows the LFV spectra applied to the full-length multiproxy data sets. The LFV spectra for data groups A–D yield statistically significant peaks on 15- to 35-year interdecadal and 50- to 130-year multidecadal/century time scales. We isolate a quarter-millennial (~ 240-year) oscillation in data group D where longer-period variability can be resolved from a secular trend. This identification is tentative, however, as less than three "cycles" are present, and the signal cannot be independently confirmed from the other data subsets. Only groups A and B can fully resolve the bidecadal signals, as the Nyquist frequency for 10-year sampling corresponds to a 20-year period. Higher-frequency signals (e.g., ENSO

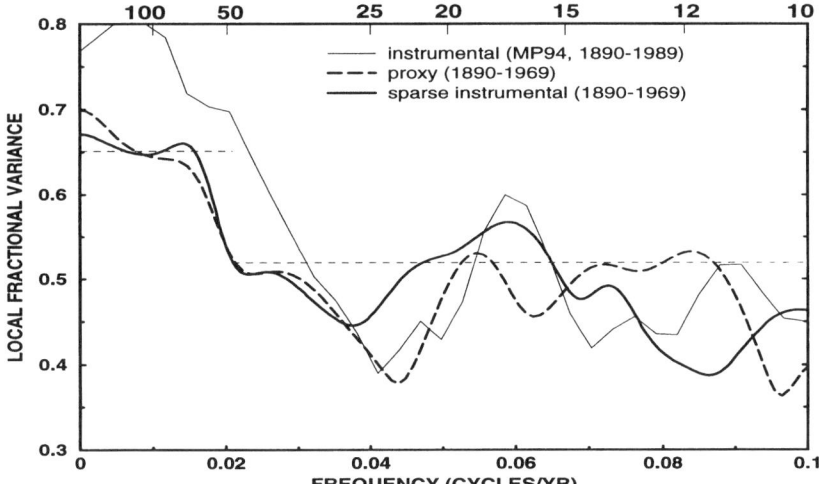

FIG. 43. LFV spectrum for (i) instrumental global temperature data of Section 4.1, (ii) proxy data set (A), (iii) instrumental data sampled similarly to the proxy network, as described in the text. The local fractional variance scale shown applies strictly only to (i), with the scales for (ii) and (iii) normalized (by factors 0.95 and 0.85, respectively) so that the quantiles of the null distribution and significance levels (which vary with the number of effective spatial degrees of freedom) approximately align at the 95% level (shown by horizontal dashed lines). Confidence limits are higher within the secular band corresponding to variability longer than about a 50-year period. [Reprinted with permission from Mann *et al* (1995b). *Nature* (London), 378, 266–270. Copyright ©1995 Macmillan Journals Limited.]

band) observed in the A and B analyses are discussed elsewhere (Bradley *et al.*, 1994). Comparison of the spatial patterns of the interdecadal and century-scale peaks reveal clear distinctions that are consistent among the data groups. We have grouped peaks of correlated variance into interdecadal and century-scale bands. The variation of these peaks among LFV spectra on different time intervals suggests signals with frequencies that drift over time.

The time-evolving nature of the amplitude and frequency of the interdecadal and century-scale signals is better examined with the evolutive MTM-SVD analysis. Window durations are chosen to be long enough to allow reasonable frequency resolution, but short enough to provide insight into the evolving character of signals. The short duration, combined with decadal resolution, precludes a meaningful evolutive analysis for data group "C." The evolutive analyses (Fig. 45) demonstrate that interdecadal oscillations, centered near 20- to 25-year periodicity, were weakly evident before 1800. The oscillations subsequently strengthen in significance and

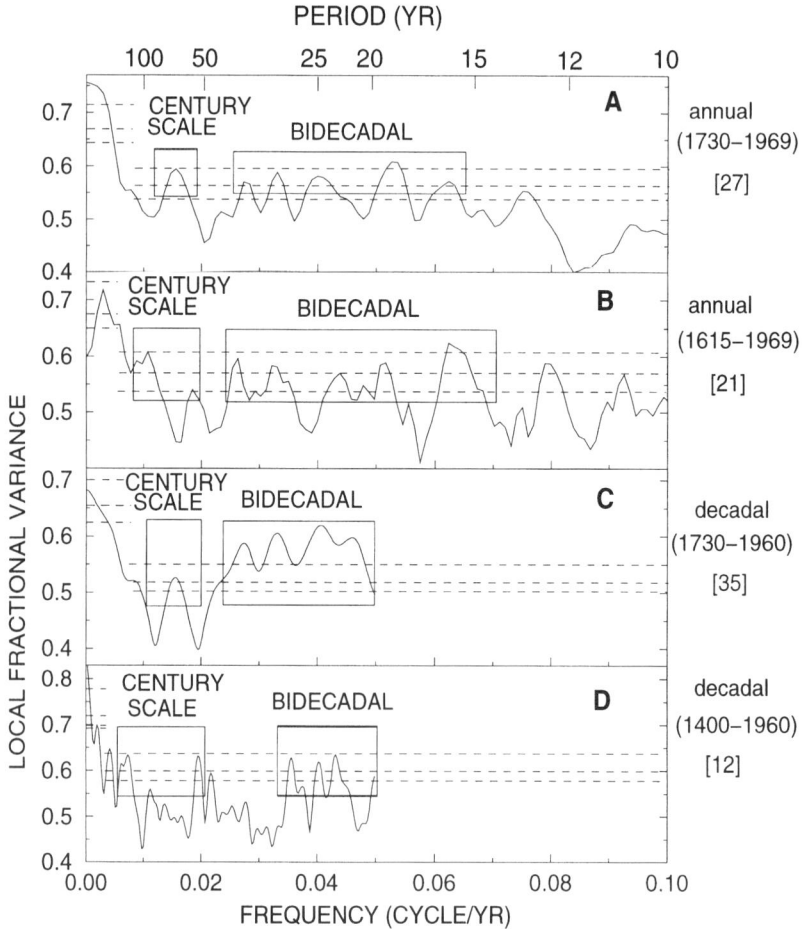

FIG. 44. LFV spectra for each of the experiments A–D. Horizontal dashed lines indicate 90%, 95%, and 99% confidence levels for significance. Significant interdecadal and century-scale peaks are indicated by surrounding boxes. [Reprinted with permission from Mann *et al* (1995b). *Nature* (London), 378, 266–270. Copyright ©1995 Macmillan Journals Limited.]

gradually increase in frequency to roughly 16- to 18-year periodicity in the final window (1869–1969), in agreement with the time scale of the interdecadal oscillation described in Sections 4.1 and 4.2. Time windows that resolve century-scale variations (200-year width) can be employed in experiment D. Before 1650, a coherent signal with roughly 50-year time scale appears intermittently. After 1650, this oscillation strengthens in its significance and drifts to a 60- to 70-year periodicity. After 1800, these

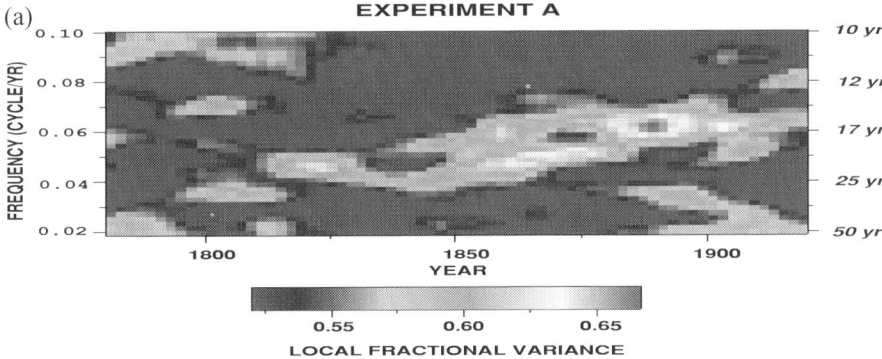
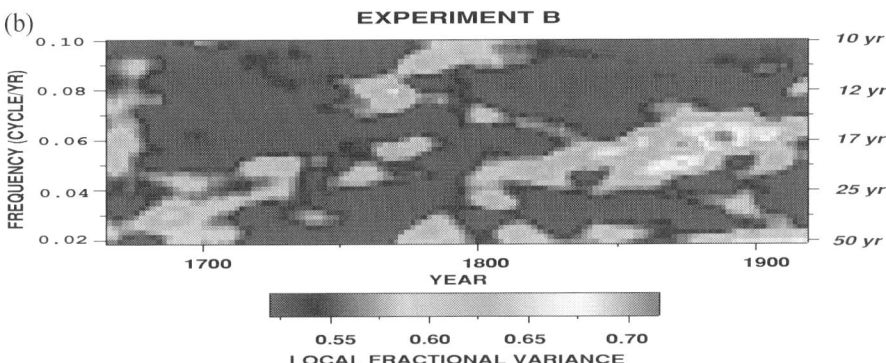
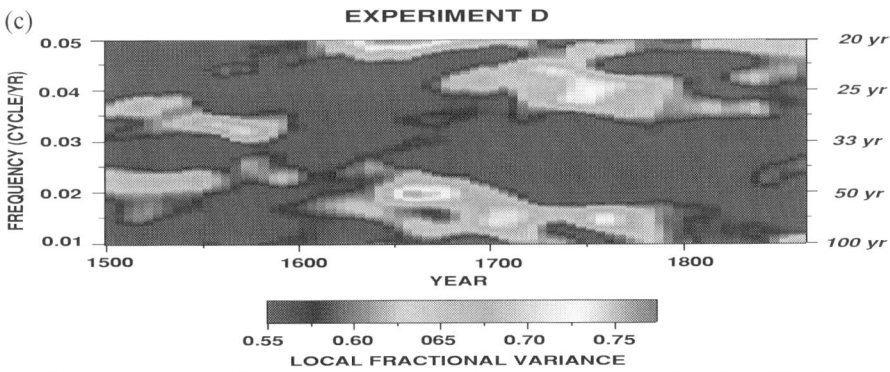

FIG. 45. Evolutive LFV spectra for (a) experiment A using a window $L = 100$. (Local fractional variance is described by a greyscale, with significance levels 90% = 0.55, 95% = 0.565, 99% = 0.60. Greyscale convention chosen so that any features clearly visible above the background are significant well above the 95% level.) (b) LFV spectrum for experiment B using a 100-year moving window (the significance levels are the same as for expriment A). (c) Experiment D using a 200-year moving window. Significance levels correspond to 0.58 (90%). 0.60 (95%), and 0.64 (99%) in the latter case. [Reprinted with permission from Mann *et al* (1995b). *Nature* (London), 378, 266–270. Copyright ©1995 Macmillan Journals Limited.]

"century-scale" oscillations appear to drift to slightly longer period, becoming indistinguishable from secular time scale variability within the confines of a 200-year moving window. Note also that the improved frequency resolution of experiment D allows for a clearer separation of the interdecadal and century-scale variability than do experiments A or B.

Signal Reconstructions

Figure 46 shows the spatial patterns of the interdecadal and century-scale signals. In the spatial reconstructions, sources of systematic bias in the temperature reconstructions may lead to unreliable phase and amplitude at isolated sites. The few records that were not originally calibrated in temperature (°C) units are calibrated using the variances of nearby instrumental gridpoint temperature data (Briffa and Jones, 1992) during the past century. Calibration of proxy data at longer periods, and various corrections that are made to proxy records (e.g., subtraction of individual long-term growth trends for dendroclimatic records), are potential sources of bias. Thus, the general regional trends that are evident in these patterns, rather than the precise response at particular sites, are most meaningful.

The interdecadal signal (Fig. 46a) exhibits variability in the tropics and subtropics that is largely in phase. Mid-latitude variations are of similar magnitude, but phase relationships are more variable, consistent with the signature of extratropical teleconnection patterns. While such patterns can be resolved by our data network only in part, the alternating pattern of phase from Greenland, to eastern, and then western mid-latitude North America (with a nodal point in central North America) is consistent with the alternating warm and cold advection of the three-lobed "Pacific North American" (PNA) pressure anomaly pattern (although the vector directions for certain sites are inconsistent with the general pattern). This PNA pattern was associated with the interdecadal signal in the instrumental climate record. Other regional proxy records (Slowey and Crowley, 1995) and analyses of continental U.S. drought reconstructions (Rajagopalan *et al.*, 1996) confirm a similar pattern of interdecadal variability in the PNA pattern. While some regional differences in relative amplitude and phase are noted with the instrumental signal (compare Fig. 24), the larger-scale features of nearly in-phase tropical/subtropical variability consistent with ENSO and an extratropical PNA-like pattern are clearly evident in the proxy climate signal. The frequency modulation of the interdecadal signal between ~ 15- and ~ 35-year-period ranges (with a recent trend towards the higher frequencies) presents a complication for the interpretation of the extratropical coupled mode of LB94 discussed in Section 4.2 ("Interde-

OSCILLATORY SPATIOTEMPORAL SIGNAL DETECTION

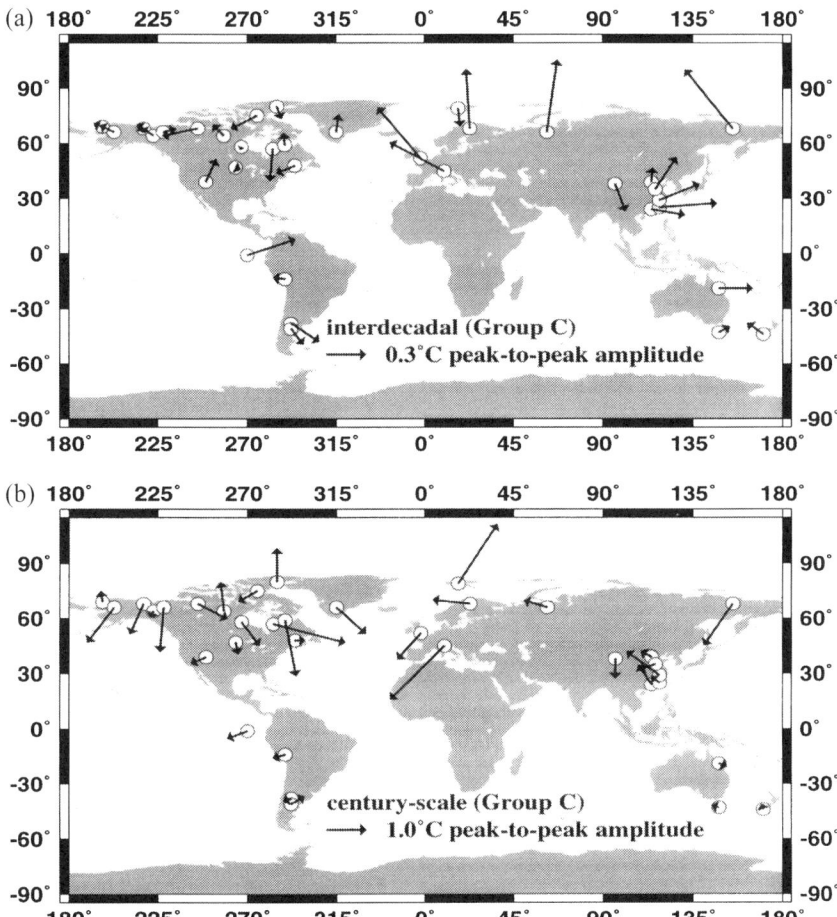

FIG. 46. (a) Spatial pattern of interdecadal oscillation based on the signal reconstruction for the ~ 24-year-period peak in experiment C (which provides the greatest spatial coverage). Conventions are similar to those introduced in Figure 22. If we define zero phase as when the tropics are generally warmest, peak warmth in northern Eurasia occurs 90° (or, roughly 6 years) later. Amplitude scale is set so that the largest arrow corresponds to the regional maximum amplitude of the oscillation of ~ 0.6°C peak-to-peak. (b) The spatial pattern for century-scale signal oscillation is reconstructed based on the 65-year-period variance peak of experiment C. Pattern maximum is a ~ 2°C peak-to-peak oscillation (e.g., central Europe). [Reprinted with permission from Mann *et al* (1995b). *Nature* (London), 378, 266–270. Copyright ©1995 Macmillan Journals Limited.]

cadal Signal"). The oscillatory time scale in the LB94 mechanism is specified by gyre spinup dynamics and should not change over time. Nor should the interdecadal mechanism of Gu and Philander (1997), which involves transit times of water mass subduction. In contrast, frequency modulation is more consistent with the behavior of a system with changing control parameters, or a nonlinear system (e.g., Lorenz, 1990; Tziperman *et al.*, 1994; Jin *et al.*, 1994). The latter connection would favor the notion that the observed interdecadal variability is associated with intrinsic low-frequency behavior of the tropical ocean–atmosphere (e.g., Graham, 1994). Thus, while the interpretation of the underlying dynamics is not obvious, the proxy data analysis does provide evidence for long-term interdecadal oscillations, and at least some suggestion of a connection with ENSO.

The spatial pattern of the "century-scale" 50- to 70-year time scale signal (Fig. 46b) exhibits high-amplitude variability largely confined to the North Atlantic and Arctic, out of phase with weaker variability in the Pacific basin. These features recall the pattern of the single multidecadal or century-scale "oscillation" described in Sections 4.1 and 4.2, although it should be noted that arbitrary phase relationships could not be determined for the century-scale signal in the instrumental record as it was confined to the secular trend frequency band. The longer history provided by the proxy data, however, allows for the recognition of a true oscillatory process on multidecadal or century time scales, with arbitrary spatial phase relationships. Aside from the Svalbard site in the boreal Atlantic, there is a tendency towards opposing, though not *opposite*, phase anomalies (i.e., $\approx 45°-135°$ phase difference) between the eastern and western margins of the North Atlantic (note that such variable phase relationships could not be captured in the short instrumental record because the multidecadal oscillation was confined to the secular frequency regime). Such a phase pattern could indicate a combination of in-phase and out-of-phase components in the two regions. An out-of-phase component could arise from differential temperature advection on either side of an alternating center of low and high pressure over the North Atlantic. Changes in northward oceanic heat transport would generate an in-phase component of basinwide warming and cooling. The combination of these effects is consistent with the coupled ocean–atmosphere model mechanisms isolated by DEL93 and discussed previously in Section 4.2, in which multidecadal (with a *slightly shorter* 40- to 60-year periodicity) variations in sea-level pressure over the North Atlantic coincide with oscillations in the thermohaline circulation. The two patterns differ somewhat, however, in that the empirical phase lag between the in-phase and out-of-phase variations across the North Atlantic suggests a temporal lag between maxima in meridional overturning heat transport to high latitudes and the most pronounced atmospheric circulation response. The simulations of DEL93 show these two responses to be more coincident. Similar empirical analyses with expanded proxy networks may better constrain the spatial relationships between the modeled and observed oscillatory signals.

Summary

MTM-SVD analysis of a globally distributed set of temperature proxy records, of several centuries duration, strengthens evidence for persistent 15- to 35-year-period "interdecadal" and 50- to 150-year "century-scale" climatic oscillations, and reveals both the spatial patterns and the temporal histories of these signals. The time-evolving amplitude and frequency of quasi-periodic signals can be examined with an "evolutive" analysis, in which the SVD analysis is applied in a moving window through the data series. The interdecadal oscillation, centered near 20- to 25-year periodicity, is weakly evident before 1800, and subsequently strengthens in significance and drifts to roughly 16- to 18-year period in the twentieth century. The century-scale mode exhibits high-amplitude variability largely confined to the North Atlantic and Arctic, out of phase with weaker variability in the Pacific basin. This behavior resembles the pattern of the single quasi-secular "oscillation" detected in gridded surface temperature and pressure data of the last 100 years.

4.4. Seasonal Cycle: Observations vs CO_2-Forced Model Simulations

Thomson (1995) showed that shifts in the phase of the annual cycle in temperature during the twentieth century are correlated with atmospheric CO_2 concentrations, and argued for an anthropogenic cause. Similar phase changes have been observed in the seasonal cycle of temperature in particular regions (Davis, 1972; Thompson, 1995) as well as shifts in the seasonality of precipitation (Bradley, 1976; Rajagopalan and Lall, 1995), streamflow (Lins and Michaels, 1994; Dettinger and Cayan, 1995), and Southern Hemisphere winds and sea-level pressure (Hurrell and Van Loon, 1994). Potential physical connections with greenhouse forcing have been suggested (Lins and Michaels, 1994), complementing the statistical correlation found by Thomson (1995). If observed changes in seasonality are consistent with an enhanced greenhouse effect, the observed trends in the seasonal cycle should resemble the simulated response of present-generation climate models to enhanced greenhouse conditions. Here, we review the comparison by Mann *et al.* (1996a) of the seasonal cycle of temperature in the Northern Hemisphere with those of simulations of (1) the Geophysical Fluid Dynamics Lab (GFDL) coupled ocean–atmosphere model (Manabe *et al.*, 1991), and (2) the NCAR Community Climate Model (CCM1) general circulation/slab ocean model (Oglesby and Saltzman, 1992).

Northern Hemisphere Average Trends

We approximate the seasonal cycle in temperature by its fundamental annual component $A(t)\cos(2\pi t + \theta(t))$, where t is time in years and the phase $\theta(t)$ and amplitude $A(t)$ can vary with time. This simple statistical model is motivated by the fact that surface temperature seasonality is determined, within a phase lag, by the yearly cycle of insolation at the top of the atmosphere in most locations. The harmonics of the annual cycle are important, however, in the tropics and in the polar latitudes of the Southern Hemisphere (see, e.g., Trenberth, 1983) and provide essential information about relationships with specific seasons (e.g., the onset of "spring" (Davis, 1972)). The departures of certain seasonal features (e.g., convective mixing in the high-latitude ocean, the termination of the monsoons, or sea-ice and snow-cover processes) from a simple annual cycle suggest that our analysis provides only a first-order estimate of more general changes in the structure of the seasonal cycle.

Using the estimated Northern Hemisphere (NH) average monthly temperature series of Jones *et al.* ("J & W," 1986—updated in Jones, 1994) with seasonal climatology intact, we estimated the variation in $\phi(t)$ and $A(t)$ of the annual cycle over the interval 1854–1990 through complex demodulation (Fig. 47).

We used three Slepian data tapers and a 10-year moving interval or "projection filter" to obtain low-variance estimates of the trends in $A(t)$ and $\phi(t)$. Through this method, phase shifts of less than one day can be resolved in monthly data (see Thomson, 1995). The calculated trends were robust as we varied the length of the moving window from 5 to 20 years. The highly variable spatial sampling (growing from $\sim 20\%$ to near-complete areal coverage during the interval under examination) may bias estimates of small changes in hemisphere-averaged quantities. To test for such bias, we analyzed alternative "frozen grid" estimates of the NH average series using gridded land air and sea surface temperature data (Jones and Briffa, 1992). These series were calculated from both (1) a "sparse" sampling of all nearly continuous gridpoint series from 1890 to 1989 (see, e.g., Mann and Park, 1994) providing 33% coverage, and (2) a "dense" sampling from 1899 to 1989 providing 53% areal coverage (shown in Fig. 48). The gross trends in the annual cycle phase and amplitude (Fig. 47) appear insensitive to the sampling of large-scale averages (see Table VI), though an unavoidable bias due to data sparseness at latitudes

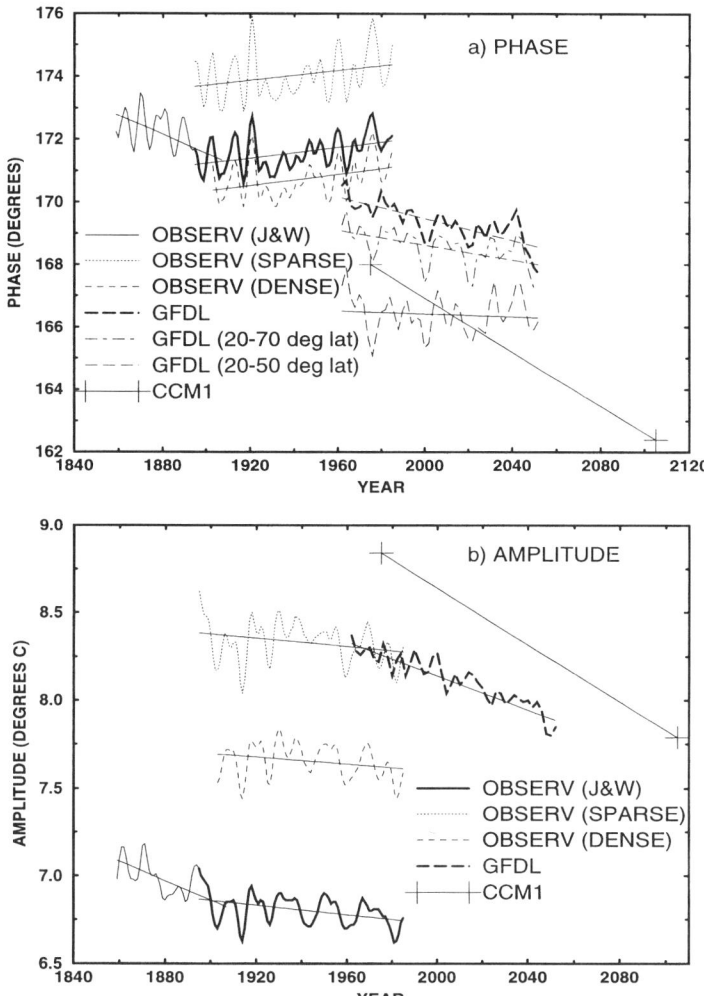

FIG. 47. (a) Phase of annual cycle in Northern Hemisphere average temperature for observations and model simulations. Best-fit linear trends (Table VI) are shown. For the observational data, results for both the J & W expanding grid, and sparse and dense "frozen-grid" estimates (see text) are indicated. For the longer J & W series, a break in slope near 1900 marks a transition from decreasing to increasing phase (latter portion shown with thicker curve). Time axis for the model defined by the actual year corresponding to the initial prescribed CO_2 level. For the CCM1 (equilibrium) experiment, only the net change has meaning. For graphical purposes, a time scale is prescribed by assuming the same 1%/year increase as in the GFDL experiment. (b) Amplitude of the annual cycle. Decreasing trends of varying magnitude are found for both model and observed data. Best-fit linear trends are shown (Table VI), with a break in slope again evident in the observations between 1890 and 1900. [From Mann and Park (1996a).]

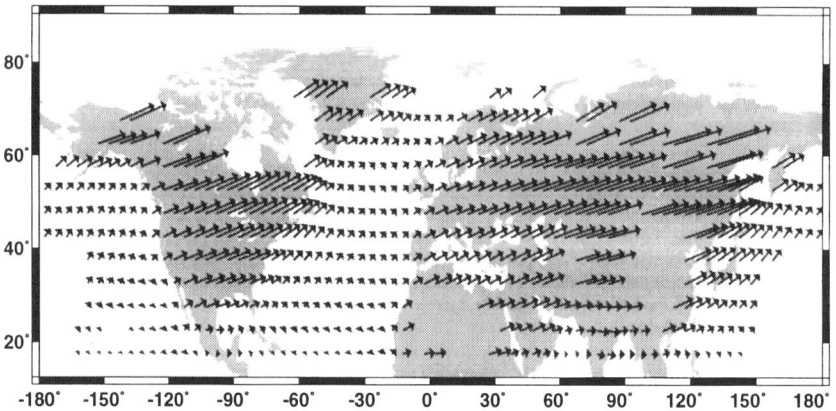

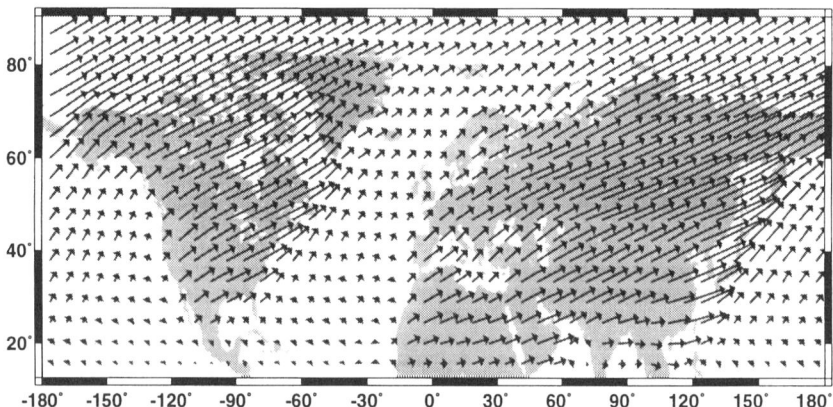

FIG. 48. Phase and amplitude of the "baseline" annual cycle in temperature for (a) observation and (b) control GFDL simulation. A phase of 0° (rightward pointing arrow) indicates a minimum temperature that coincides with minimum insolation (Dec. 22) in the Northern Hemisphere. A 30° counterclockwise rotation indicates a 1-month phase delay of minimum temperature relative to the insolation minimum. [From Mann and Park (1996a).]

TABLE VI LINEAR TRENDS IN PHASE AND AMPLITUDE OF NORTHERN HEMISPHERE AVERAGE ANNUAL TEMPERATURE CYCLE FOR MODEL AND OBSERVATIONS [FROM MANN AND PARK (1996A).]

Series	$\Delta\phi(°)$	Significance	ΔA	Significance
Observ (J & W: 1854–1904)	−1.4	> 5.5σ	−0.55	> 6σ
Observ (J & W: 1899–1989)	0.79	> 4.5σ	−0.13	> 4σ
Observ (sparse)	0.71	> 3σ	−0.12	> 3σ
Observ (dense)	0.83	> 3.5σ	−0.10	> 2.5σ
GFDL (CO_2 increase)	−1.7	> 7.5σ	−0.48	> 16σ
GFDL (20–70°)	−1.2	> 4σ	−0.50	> 10σ
GFDL (20–50°)	−0.2	< σ	−0.04	< σ
GFDL (control)	0.5	> 1.5σ	+0.02	< σ
CCM1 (460 ppm–330 ppm)	−5.6		−1.05	

poleward of 70°N may exist. The baseline annual cycle varies with the mixture of land, ocean, and high-latitude gridpoints due to important regional variations in phase and amplitude.

Trends towards an advanced phase (i.e., earlier seasonal transitions) are significant at better than 2.5σ in each of the three data schemes (Table VI) based on jackknife uncertainties, taking serial correlation into account. Such significance does not alone indicate a causal connection with greenhouse-related warming, as it could result, for example, from the enhanced century-scale *natural* variability that is evident in both observations (Mann and Park, 1994; Schlesinger and Ramankutty, 1994; Mann et al., 1995b) and modeling studies (Delworth et al., 1993). If an opposing trend towards delayed phase due to orbital precession is adopted as a null hypothesis (Thomson, 1995), the above trends become more significant.

A significant decreasing trend in $A(t)$ is also found for each data-weighting scheme (Fig. 47b). A break in the slope between 1884 and 1895 is significant at the $p = 0.01$ level. Lean et al. (1995) suggest an increasing trend in solar irradiance beginning in the early twentieth century. This trend could counteract an even greater decrease in $A(t)$ that might arise from global warming and associated ice-albedo feedback, potentially explaining the break in slope. A connection between decreasing $A(t)$ and decreased winter ice cover is suggested by the model responses to greenhouse forcing.

We analyzed for comparison both (i) the change in the CCM1 climatological annual cycle between 330 ppm and 460 ppm CO_2 level equilibrations (see Oglesby and Saltzman, 1992; Marshall et al., 1995) and (ii) 100-year simulations of the GFDL coupled model (e.g., Manabe et al., 1991) (a) with a gradual (1%/year) CO_2 increase and (b) with fixed present-day CO_2. Both models exhibit a significant annual cycle response to greenhouse forcing (Table VI). Decreased amplitude of the annual cycle under CO_2-enhanced conditions is consistent with the observations. The time axis for the transient GFDL model simulations should be interpreted quite loosely, as the imposed forcing in these simulations is highly idealized and does not realistically mimic changes in observed greenhouse gas concentrations. The trend in phase for the models, however, is opposite to that observed, exhibiting a delay, rather than an advance, of the seasons. The magnitude and significance of the trends in the enhanced-greenhouse GFDL simulation diminishes if high- and low-latitude regions, poorly sampled by the observational data, are excluded (Fig. 47a), but no latitude band exhibits the phase *advance* found in the observations. The control GFDL simulation, like the observations, exhibits a marginally significant advance in phase (Table VI), perhaps associated with organized century-scale variability (Delworth *et al.*, 1993).

Spatial Patterns

To reconstruct the spatial patterns of the climatological annual cycle, we used a multivariate generalization (Mann and Park, 1994; Mann *et al.*, 1995a, b) of the complex demodulation procedure used by Thomson (1995). The climatological seasonal cycle in the control GFDL simulation resembles quite closely that for the "dense" observational temperature sampling (Fig. 49). It should, however, be noted that this is partly due to seasonally specific flux corrections that are imposed in the model at the ocean surface (Manabe *et al.*, 1991). These climatological flux corrections, furthermore, may suppress the tendency for the annual cycle in the model to depart from its baseline state. The annual cycle over continents is delayed by ~ 1 month relative to the insolation cycle, due to the thermal capacity of land, continental snow cover, and other climatic factors; see Trenberth (1983) for an overview. The greater thermal capacity of the oceans leads to a greater delay (typically, 2 months) and a smaller annual cycle amplitude. Land areas strongly influenced by the oceans experience a more maritime annual cycle. Winter sea ice insulates the ocean surface from the mixed layer, exposing some oceanic regions to cold continental outbreaks. This can lead to a more "continental" seasonal cycle in the high-latitude oceans. Changes in the annual cycle could thus arise from many influences. The climatological annual cycle of the CCM1 (not shown) reproduces the observations less well. CCM1 predicts an oceanic phase lag that is typically ~ 1 month too large because the slab ocean is a poor approximation to the true mixed layer.

GFDL Annual Cycle
a) change in phase

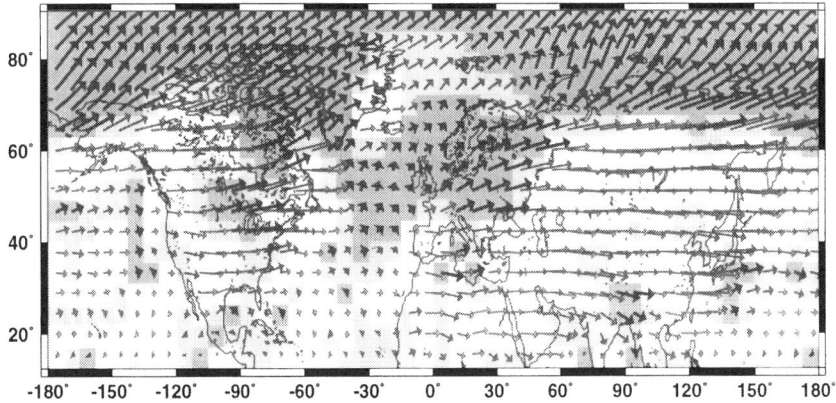

GFDL Annual Cycle
b) change in amplitude

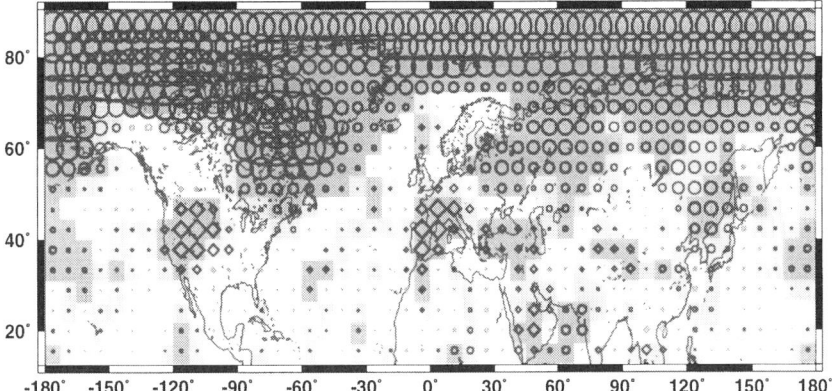

FIG. 49. The linear trend of the annual cycle for the enhanced-greenhouse GFDL simulation (a) phase and (b) amplitude. Size of arrows scales the average amplitude of the annual cycle, while the direction indicates relative delay or advance of the annual cycle. A rightward arrow indicates no change in phase. Clockwise and counterclockwise rotation indicate phase advances and delays, respectively. Significance of trends is indicated in terms of the ratio of the phase shift to its jackknife uncertainty estimate. Boldface symbols/darkest shading indicate nonzero phase and amplitude shifts at the 2σ level, thin black symbols/medium shading indicate nonzero shifts at the 1σ level, and gray symbols/no shading indicate shifts within 1σ of zero. [From Mann and Park (1996a).]

To determine the spatial pattern of annual cycle trends in the GFDL simulations, we used the multivariate procedure described above to isolate the average annual cycle in successive 10-year intervals. We regressed the long-term trends in $\theta(t)$ and $A(t)$ on a gridpoint-by-gridpoint basis, calculating jackknife uncertainties from the decadal averages. The spatial pattern of the CCM1 response (not shown) was estimated by differencing the 460-ppm and 330-ppm equilibrium climatologies.

The dominant response in both the CCM1 and GFDL models to increased CO_2 is one of substantial phase delays and amplitude decreases in high-latitude oceanic regions. We interpret phase trend as arising from decreased winter sea-ice cover and greater exposure of the surface to the ocean's mixed layer and its delayed thermal cycle. The amplitude trend is consistent with a strong positive ice-albedo feedback from reduced winter ice-cover. The close similarity of the primary response in these two very different model experiments suggests a consistent dynamical mechanism. Nonetheless, a more spatially complex trend pattern in the GFDL coupled model (Fig. 49) suggests other potential regional effects. Marginally significant phase advances, for example, are found in south central and eastern Asia. In the western U.S., the phase advance and amplitude increase suggest decreased maritime influence. The significance of these features, however, is comparable to that observed in the control experiment, suggesting that they may be associated with the model's natural century-scale variability rather than with a greenhouse response, or perhaps with some combination of these effects.

Observed amplitude trends (Fig. 50) are $-2.4°C < \delta A < +1.0°C$. Phase advances and delays of 3–7° (i.e., 3 to 7 days) are common. The largest δA is along the western margins of Greenland, where significant winter warming has occurred during the last century (Jones and Briffa, 1992). Here, we also find the most significant trend towards a delayed (~ 8 days) annual cycle in the Northern Hemisphere consistent with the model-simulated signature of greenhouse-related decreases in high-latitude sea ice. In contrast, the phase of the annual cycle has advanced along the eastern margins of Greenland, where a long-term winter cooling trend is observed (Jones and Briffa, 1992). This cooling appears to be associated with organized century-scale variability in the North Atlantic (see Mann and Park, 1994; Schlesinger and Ramankutty, 1994; Mann et al., 1995b), which could explain why the signature of greenhouse forcing is masked in this region. The annual cycle amplitude decreases in this location because winter cooling is offset by even greater summer cooling. A broad region of significant trends in annual cycle phase and amplitude is found in the extreme southwestern U.S. and offshore in the subtropical Pacific. This may be related to secular changes in the El Niño/Southern Oscillation

Observational Annual Cycle
a) change in phase

↗ 15°C annual cycle
4-day phase delay

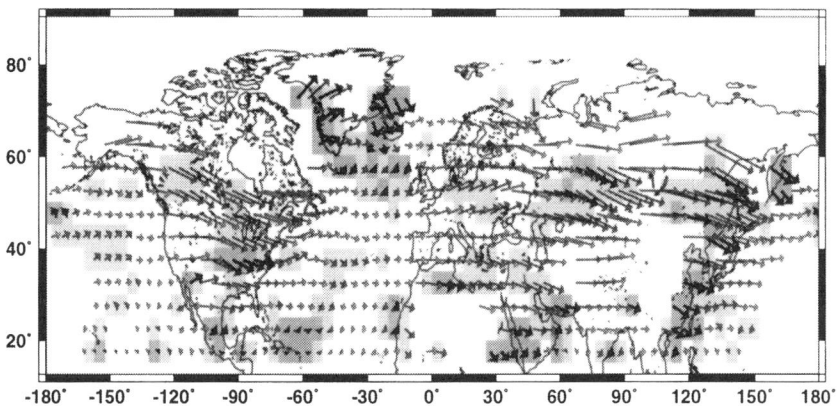

Observational Annual Cycle
b) change in amplitude

◇ +1°C per century
○ -1°C per century

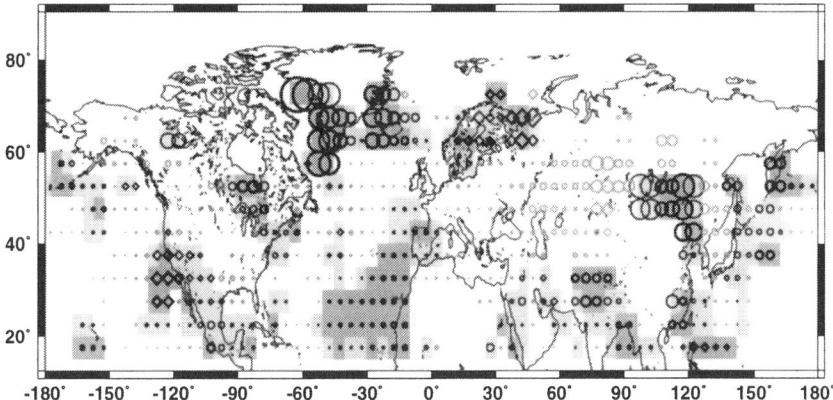

FIG. 50. The linear trend in (a) phase and (b) amplitude of the annual cycle for the "dense" observational network of gridded land air and sea surface temperature data from 1899–1990 discussed in the text. Significance of trends indicated as in Figure 49. [From Mann and Park 1996a.]

(ENSO) and associated changes in patterns of summer coastal upwelling (e.g., Trenberth and Hurrell, 1994; Graham, 1995).

A combination of phase advances and amplitude decreases over mid-latitude continental interiors is consistent with an earlier snowmelt and runoff (Lins and Michaels, 1994; Dettinger and Cayan, 1995; Groisman *et al.*, 1994) that may be related to greenhouse warming (Lins and Michaels, 1994; Groisman *et al.*, 1994). Few other locations in the Northern Hemisphere exhibit a consistent, readily interpretable annual cycle response. The constructive addition of trends in continental-interior regions is primarily responsible for the average ~ 1-day phase advance for the Northern Hemisphere.

Discussion and Summary

Both observations and model responses to greenhouse forcing show a trend towards decreased amplitude of the seasonal cycle in NH average temperatures. The simulations suggest that these amplitude decreases may result from ice-albedo feedback. It is here, however, that the agreement ends; the observed and model-predicted trends in the phase of the seasonal cycle show little similarity.

If, as the models simulate, the dominant influence on annual-cycle amplitude $A(t)$ and phase $\phi(t)$ stems from high-latitude sea-ice decreases, the signature of greenhouse warming is scarcely evident in the observational data, which lack widespread high-latitude sampling. The trend in Western Greenland, the highest-latitude region in the observations, does nonetheless resemble model simulations. Some of these discrepancies could be due to the influence of sulfate aerosols, which may have masked the effects of enhanced greenhouse gases in certain regions (see, e.g., IPCC, 1996, Chapter 8) and are not included as forcings in the simulations analyzed here. It should be noted, however, that the physical mechanisms are not yet well understood (Hansen *et al.*, 1997). It is possible that observed trends in phase, largely influenced by mid-latitude continental interiors, do not arise from greenhouse warming, but rather, at least in part, from natural variability. Such a notion is reinforced by the fact that marginally significant trends are found in the control GFDL annual cycle, presumably due to organized century-scale internal variability.

If, on the other hand, the observed variation in the seasonal cycle truly represents a "fingerprint" of greenhouse warming, the GFDL and CCM1 models do not appear capable of capturing the detailed responses of the seasonal cycle to greenhouse forcing. In particular, if the phase advances that result from the behavior in continental interiors are not only statistically (as Thomson (1995) suggests), but in fact, causally related to greenhouse forcing, the predicted behavior of the models in these regions would appear to be flawed. Deficiencies in certain aspects of the models (e.g., land surface parameterizations) could plausibly be at fault in such a scenario. The absence of an ENSO of realistic amplitude is also a potential shortcoming of model-predicted changes in seasonality, as some of the observed trends appear to show connections with ENSO.

It is possible, probably likely, that the observed trends in the seasonal cycle represent a combination of internal variability, enhanced greenhouse effects, and external forcings. Various alternative scenarios are difficult to resolve, owing to limitations in the observational data and potential shortcomings in the models' descriptions of certain climate processes. The latter limitation may largely be overcome in newer-generation climate models. Discrepancies between the observed and the model-predicted trends must be resolved before a compelling connection can be drawn between twentieth-century changes in the behavior of the annual cycle in temperature and anthropogenic forcing of the climate.

5. Conclusion

We have shown that traditional signal-detection techniques suffer a number of weaknesses or limitations in the detection and reconstruction of irregular spatiotemporal oscillatory signals immersed in colored noise. A methodology for signal detection and reconstruction of such signals—the MTM-SVD methodology—is offered as an alternative technique which avoids most of these problems and provides an efficient exploratory method for climate signal detection. The associated signal-detection parameter—the LFV spectrum—yields the correct null distribution for a very general class of spatiotemporal climate noise processes, and the correct inferences when signals are present. The methodology allows for a faithful reconstruction of the arbitrary spatiotemporal patterns of narrow-band signals immersed in spatially correlated noise. Furthermore, the results of the MTM-SVD approach are robust to the temporal and spatial sampling inhomogeneities that are common in actual climate data.

Applied to observational climate data, the MTM-SVD analysis yields insight into secular trends, low-frequency and high-frequency quasi-oscillatory variations in the climate system. The dominant mode of secular variation during the last century is a long-term global warming trend associated with some anomalous atmospheric circulation patterns that show similarity to the modeled response of the climate to increased greenhouse gases. Nonetheless, a substantial $\sim$ 70-year "century-scale" or "multidecadal" secular variation suggestive of longer-term oscillatory behavior is superimposed on this trend in both the global temperature and joint Northern Hemisphere temperature/SLP fields, exhibiting substantial SLP and temperature variations in the North Atlantic, and displaying a notable high-latitude signature in the temperature field. The analysis of multiple centuries of proxy data suggests that 50- to 100-year internal oscillations with similar features persist over several centuries. Similar oscillatory signals were attributed to variability in the thermohaline circulation and possible coupled ocean–atmosphere processes in recent model simulation studies. An analysis of the seasonal cycle in surface temperature in the observations and long coupled model integrations suggests the impact of both anthropogenic forcing and multidecadal oscillatory variations on seasonality in surface temperature.

An interdecadal 16- to 18-year climate signal is clearly evident in the instrumental climatic fields analyzed, and appears at some level to be consistent with a mechanism involving gyre spinup and mid-latitude ocean–atmosphere interaction, which has also been predicted in a recent coupled ocean–atmosphere simulation. A connection with decadal-scale ENSO variability, suggested both by correlations with standard ENSO indices and by ENSO-like teleconnections in temperature and atmospheric circulation, suggest a complexity to the signal which has not been well explained. Analysis of long-term proxy data demonstrates evidence for the persistent, if intermittent, nature of this cycle over many centuries. The frequency modulation of the interdecadal signal evident in these longer-term data seems to complicate the interpretation in terms of any simple linear dynamical mechanism. Weaker quasi-decadal oscillations are evident, but with statistical confidence. Our analysis also substantiates the importance of oscillatory behavior on interannual time scales associated with ENSO and quasi-biennial oscillations, and provides insight into the long-term changes in such relatively high-frequency climatic processes.

Acknowledgments

M.E.M. acknowledges support from the U.S. Department of Energy through the Alexander Hollaender Distinguished Postdoctoral Research Fellowship program. This work was originally supported by the NSF Climate Dynamics Program, Grant ATM-9222592.

The authors greatly appreciate the thoughtful and constructive review provided by John Lanzante. Helpful comments were also provided by David Parker and Barry Saltzman. Valuable feedback and constructive criticism during various stages of development and application of this research is acknowledged from many individuals, including Myles Allen, Philip Bogden, Michael Ghil, Yochanan Kushnir, Upmanu Lall, Robert Livezey, Balaji Rajagopalan, Ronald Smith, and David Thomson.

References

Allen, M. R., and Robertson, A. W. (1996). Distinguishing modulated oscillations from coloured noise in multivariate datasets. *Clim. Dyn.* **12**, 775–784.
Allen, M. R., and Smith, L. A. (1994). Investigating the origins and significance of low-frequency modes of climatic variability. *Geophys. Res. Lett.* **21**, 883–886.
Allen, M. R., Read, P. L., and Smith, L. A. (1992). Temperature time series. *Nature* **355**, 686.
Angell, J. K. (1990). Variation in global tropospheric temperature after adjustment for the El Niño influence, 1958–1989. *Geophys. Res. Lett.* **17**, 1093–1096.
Barnett, T. P. (1983). Interaction of the monsoon and Pacific tradewind systems at interannual time scales. Part I: The equatorial zone, *Mon. Weather Rev.* **111**, 756–773.
Barnett, T. P. (1991). The interaction of multiple time scales in the tropical climate system. *J. Climate* **4**, 269–281.
Barnett, T. P., Santer, B., Jones, P. D, and Bradley, R. S. (1996). Estimates of low frequency natural variability in near-surface air temperature. *Holocene* **6**, 255–263.
Barnston, A. G., and Livezey, R. E. (1987). Classification, seasonality and persistence of low-frequency atmospheric circulation patterns. *Mon. Weather Rev.* **115**, 1083–1126.
Barnston, A. G., Livezey, R. E., and Halpert, M. S. (1991). Modulation of Southern Oscillation–northern hemisphere mid-winter climate relationships by the QBO. *J. Climate* **4**, 203–227.
Bottomley, M., Folland, C. K., Hsiung, J., Newell, R. E., and Parker, D. E. (1990). Global ocean surface temperature atlas (GOSTA). Joint Meteorological Office and Massachusetts Institute of Technology Project, U.K. Depts. of the Environment and Energy, HMSO, London.
Bradley, R. S. (1976). "Precipitation History of the Rocky Mountain States." Westview Press, Boulder, CO.
Bradley, R. S. and Jones, P. D. (1993). "Little Ice Age" summer temperature variations: Their nature and relevance to recent global warming trends. *Holocene* **3**, 367–376.
Bradley, R. S., Mann, M. E., and Park, J. (1994). A spatiotemporal analysis of ENSO variability based on globally distributed instrumental and proxy temperature data. *Eos Suppl.* **75**, 383.
Bretheron, C. S., Smith, C., and Wallace, J. M. (1992). An intercomparison of methods for finding coupled patterns in climate data. *J. Climate* **5**, 541–560.
Briffa, K. R. and Jones P. D. (1993). Surface air temperature variations during the 20th century: Part 2—Implications for large-scale high-frequency paleoclimate studies. *Holocene* **3**, 82–92.
Brillinger, D. (1981). "Time Series—Data Analysis and Theory." McGraw-Hill, New York, 1981.
Burroughs, W. J. (1992). "Weather Cycles: Real or Imaginary?" Cambridge Univ. Press, Cambridge, UK, 1992.
Bursor, G. (1993). Complex principal oscillation pattern analysis. *J. Climate* **6**, 1972–1986.
Cai, W., and Godfrey, S. J. (1995). Surface heat flux parameterizations and the variability of the thermohaline circulation. *J. Geophys. Res.* **100**, 10679–10692.
Cane, M. A., and Zebiak, S. E. (1987). Prediction of El Niño events using a physical model. *In* "Atmospheric and Oceanic Variability" (H. Cattle, ed.). Royal Meteorological Society Press, Bracknell, UK pp. 153–182.
Cane, M. A., Clement, A. C., Kaplan, A., Kushnir, Y., Pozdnyakov, D., Seager, R., Zebiak, S. E., and Murtugudde, R. (1997). Twentieth-century sea surface temperature trends. *Science* **275**, 957–960.

Cane, M. A., Zebiak, S. E., and Dolan, S. C. (1986). Experimental forecasts of El Niño. *Nature* **321,** 827–832.

Cayan, D. R. (1992a). Latent and sensible heat flux anomalies over the northern oceans: The connection to monthly atmospheric circulation. *J. Climate* **5,** 354–369.

Cayan, D. R. (1992b). Latent and sensible heat flux anomalies over the northern oceans: Driving the sea surface temperature. *J. Phys. Oceanogr.* **22,** 859–881.

Chang, P., Link, J., and Li, H. (1997). A decadal climate variation in the tropical Atlantic Ocean from thermodynamic air–sea interactions. *Nature* **385,** 516–518.

Chen, F., and Ghil, M. (1995). Interdecadal variability of the thermohaline circulation and high-latitude surface fluxes. *J. Phys. Oceanogr.* **25,** 2547–2568.

Chen, F., and Ghil, M. (1996). Interdecadal variability in a hybrid coupled ocean–atmosphere model. *J. Phys. Oceanogr.* **26,** 1561–1578.

Cheng, X., and Dunkerton, T. J. (1995). Rotation of spatial patterns derived from singular value decomposition analysis. *J. Climate* **8,** 2631–2643.

Cherry, S. (1997). Some comments on singular value decomposition analysis. *J. Climate* **10,** 1759–1766.

Cole, J. E., Fairbanks, R. G., and Shen, G. T. (1993). Recent variability in the Southern Oscillation: Isotopic results from a Tarawa atoll coral. *Science* **260,** 1790–1793.

Currie, R. G., and O'Brien, D. P. (1992). Deterministic signals in USA precipitation records; II. *Int. J. Climatol.* **12,** 281–304.

Darby, M. S., and Mysak, L. A. (1993). A boolean delay equation model of an interdecadal Arctic climate cycle. *Clim. Dyn.* **8,** 241–246.

Davis, N. E. (1972). The variability of the onset of spring in Britain. *Quart. J. Roy. Meteorol. Soc.* **418,** 763.

Delworth, T., Manabe, S., and Stouffer, R. J. (1993). Interdecadal variations of the thermohaline circulation in a coupled ocean–atmosphere model. *J. Climate* **6,** 1993–2011.

Delworth, T. D., Manabe, S., and Stouffer, R. J. (1997). Multidecadal climate variability in the Greenland Sea and surrounding regions: A coupled model simulation. *Geophys. Res. Lett.* **24,** 257–260.

Deser, C., and Blackmon, M. (1993). Surface climate variations over the North Atlantic ocean during winter: 1900–1989. *J. Climate* **6,** 1743–1753.

Dettinger, M. D., and Cayan, D. R. (1995). Large-scale atmospheric forcing of recent trends toward early snowmelt runoff in California. *J. Climate* **8,** 606.

Dettinger, M. D., Ghil, M., and Keppenne, C. L. (1995). Interannual and interdecadal variability in United States surface-air temperatures, 1910–1987. *Clim. Change* **31,** 36–66.

Diaz, H., and Pulwarty, R. S. (1994). An analysis of the time scales of variability in centuries-long ENSO-sensitive records in the last 1000 years. *Clim. Change* **26,** 317–342.

Dickey, J. O., Marcus, S. L., and Hide, R. (1992). Global propagation of interannual fluctuations in atmospheric angular momentum. *Nature* **357,** 484–488.

Dickson, B. (1997). From the Labrador Sea to global change. *Nature* **386,** 649–650.

Dunbar, R. B., Wellington, G. M., Colgan, M. W., and Glynn, P. W. (1994). Eastern Pacific sea surface temperature since 1600 A.D.: The δ^{18} record of climate variability in Galapagos corals. *Paleocn.* **9,** 291–315.

Efron, B. (1990). "The Jackknife, the Bootstrap and Other Resampling Plans." SIAM, Philadelphia.

Folland, C. K., Palmer, T. N., and Parker, D. E. (1986). Sahel rainfall and worldwide sea temperatures. *Nature* **320,** 602–606.

Folland, C. K., Parker D. E., and Kates, F. E. (1984). Worldwide marine temperature fluctuations 1856–1981. *Nature* **310,** 670–673.

Fraedrich, K., Bantzer, C., and Burkhardt, U. (1993). Winter climate anomalies in Europe and their associated circulation at 500 hPa. *Clim. Dyn.* **8**, 161–175.
Friis-Christensen, E., and Lassen, K. (1991). Length of the solar cycle: An indicator of solar activity closely associated with climate. *Science* **254**, 698–700.
Ghil, M., and Vautard, R. (1991). Interdecadal oscillations and the warming trend in global temperature time series. *Nature* **350**, 324–327.
Ghil, M., and Yiou, P. (1996). Spectral methods: What they can and cannot do for climatic time series. *In* "Decadal Climate Variability." Springer-Verlag, Berlin/New York, pp. 446–482.
Gilman, D. L., Fuglister, F. J., and Mitchell, Jr., J. M. (1963). On the power spectrum of "red noise." *J. Atmospheric Sci.* **20**, 182–184.
Gordon, A. L., Zebiak, S. E., and Bryan, K. (1992). Climate variability and the Atlantic Ocean, *Eos Trans., AGU* **73**, 161–165.
Graham, N. E. (1994). Decadal-scale climate variability in the tropical and North Pacific during the 1970s and 1980s: Observations and model results. *Clim. Dyn.* **10**, 135–162.
Graham, N. E. (1995). Simulation of recent global temperature trends. *Science* **267**, 666–671.
Graham, N. E., Michaelsen, J., and Barnett, T. P. (1987). An investigation of the El Niño-Southern Oscillation cycle with statistical model. Predictor field characteristics. *J. Geophys. Res.* **92**, 14251–14270.
Greatbatch, R. J., and Zhang, S. (1995). An interdecadal oscillation in an idealized ocean basin forced by constant heat flux. *J. Climate* **8**, 81–91.
Griffies, S. M., and Bryan, K. (1997). Predictability of North Atlantic multidecadal climate variability. *Nature* **275**, 181–184.
Griffies, S. M., and Tziperman, E. (1995). A linear thermohaline oscillator driven by stochastic atmospheric forcing. *J. Climate* **8**, 2440–2453.
Groisman, P. Ya, Karl, T. R., and Knight, R. W. (1994). Observed impact of snow cover on the heat balance and the rise of continental spring temperatures. *Science* **263**, 198.
Gu, D., and Philander, S. G. H. (1997). Interdecadal climate fluctuations that depend on exchanges between the tropics and extratropics. *Science* **275**, 805–807.
Halpert, M. S., and Ropelewski, C. F. (1992). Surface temperature patterns associated with the Southern Oscillation. *J. Climate* **5**, 577–593.
Hansen, J., Sato, M., and Ruedy, R. (1997). The missing climate forcing, *Philos. Trans. Roy. Soc. London Ser. B* **352**, 231–240.
Hasselmann, K. (1976). Stochastic climate models, Part I. Theory. *Tellus* **28**, 473–478.
Hasselmann, K. (1988). PIPs and POPs: The reduction of complex dynamical systems using principal interaction and oscillation patterns. *J. Geophys. Res.* **93**, 11015–11021.
Horel, J. D., and Wallace, J. M. (1981). Planetary-scale atmospheric phenomena associated with the Southern Oscillation. *Mon Weather Rev.* **109**, 813–829.
Houghton, R., and Tourre, Y. (1992). Characteristics of low-frequency sea-surface temperature fluctuations in the tropical Atlantic. *J. Climate* **5**, 765–771.
Huang, R. X. (1993). Real freshwater flux as a natural boundary condition for the salinity balance and thermohaline circulation forced by evaporation and precipitation. *J. Phys. Oceanogr.* **23**, 2428–2446.
Hughes, M., and Diaz, H. F. (1994). Was there a "medieval warm period," and if so, where and when? *Clim. Change* **26**, 109–142.
Hurrell, J. (1995). Decadal trends in the North Atlantic Oscillation and relationship to regional temperature and precipitation. *Science* **269**, 676–679.
Hurrell, J. W., and van Loon, H. (1994). A modulation of the atmospheric annual cycle in the Southern Hemisphere. *Tellus* **46A**, 325–338.

IPCC. (1996). "Climate Change 1995, The Science of Climate Change" (J. T. Houghton, L. G. Meira Filho, B. A. Callender, N. Harris, A. Kattenberg, and K. Maskell, eds.). Cambridge Univ. Press, Cambridge, UK, 1996.

Jacoby, G. C., and D'arrigo, R. (1989). *Clim. Change* **14**, 39–59.

Jin, F. F, Neelin, J. D., and Ghil, M. (1994). El Niño on the devil's staircase: Annual subharmonic steps to chaos. *Science* **264**, 70–72.

Jones, P. D. (1989). The influence of ENSO on global temperatures. *Clim. Monit.* **17**, 80–89.

Jones, P. D. (1994). Hemispheric surface temperature variations: A reanalysis and an update to 1993. *J. Climate* **7**, 1794–1802.

Jones, P. D., and Briffa, K. R. (1992). Global surface air temperature variations during the 20th century: Part 1—Spatial, temporal and seasonal details. *Holocene* **1**, 165–179.

Jones, P. D., Raper, S. C., Bradley, R. S., Diaz, H. F., Kelly, P. M., and Wigley, T. M. (1986). Northern hemisphere surface air temperature variations. *J. Clim. Appl. Meteorol.* **25**, 161–179.

Kaplan, A., Cane, M. A., Kushnir, Y., Clement, A. C., Blumenthal, M. B., and Rajagopalan, B. (1999). Analyses of global sea surface temperature 1865–1991. *J. Geophys. Res.*, in press.

Keppenne, C. L., and Ghil, M. (1992). Adaptive filtering and prediction of the Southern Oscillation index. *J. Geophys. Res.* **97**, 20449–20454.

Keppenne, C. L., and Ghil, M. (1993). Adaptive filtering and prediction of noisy multivariate signals: An application to atmospheric angular momentum. *Internat. J. Bifurc. and Chaos* **3**, 625–634.

Kim, K. Y., and North, G. R. (1991). Surface temperature fluctuations in a stochastic climate model. *J. Geophys. Res.* **96**, 18573–18580.

Koch, D., and Mann, M. E. (1996). Spatial and temporal variability of 7Be surface concentrations. *Tellus.* **48B**, 387–396.

Kuo, C., Lindberg, C., and Thomson, D. J. (1990). Coherence established between atmospheric carbon dioxide and global temperature. *Nature* **343**, 709–713.

Kurgansky, M. V., Dethloff, K., Pisnichenko, I. A., Gernandt, H., Chmielewski, F.-M., and Jansen, W. (1996). Long-term climate variability in a simple nonlinear atmospheric model. *J. Geophys. Res.* **101D**, 4299–4314.

Kushnir, Y. (1994). Interdecadal variations in North Atlantic sea surface temperature and associated atmospheric conditions. *J. Climate* **7**, 141–157.

Kutzbach, J. E., and Bryson, R. A. (1974). Variance spectrum of Holocene climatic fluctuations in the North Atlantic Sector. *J. Atmospheric Sci.* **31**, 1958–1963.

Labitzke, K., and van Loon, H. (1988). Associations between the 11-year solar cycle, the QBO, and the atmosphere. Part I: The troposphere and stratosphere in the northern hemisphere in winter. *J. Atmospheric Terrestrial Phys.* **50**, 197–206.

Lall, U., and Mann, M. (1995). The Great Salt Lake: A barometer of low-frequency climatic variability. *Water Resourc. Res.* **31**, 2503–2515.

Lamb, P. J., and Peppler, R. A. (1987). North Atlantic oscillation: Concept and an application, *Bull. Am. Meteorol. Soc.* **68**, 1218–1225.

Lanzante, J. R. (1990). The leading modes of 10–30-day variability in the extratropics of the Northern Hemisphere during the cold season. *J. Atmospheric Sci.* **47**, 2115–2140.

Latif, M., and Barnett, T. P. (1994). Causes of decadal climate variability over the North Pacific and North America. *Science* **266**, 634–637.

Lean, J., Beer, J., and Bradley, R. S. (1995). Comparison of proxy records of climate change and solar forcing. *Geophys. Res. Lett.* **22**, 3195–3198.

Lilly, J., and Park, J. (1995). Multiwavelet spectral and polarization analysis of seismic records. *Geophys. J. Internat.* **122**, 1001–1021.

Lins, H. F., and Michaels, P. J. (1994). Increasing U.S. streamflow linked to greenhouse forcing. *Eos* **75,** 281–285.

Linsley, B. K., Dunbar, R. B., Wellington, G. M., and Mucciarone, D. A. (1994). A coral-based reconstruction of intertropical convergence zone variability over Central America since 1707. *J. Geophys. Res.* **99,** 9977–9994.

Liu, Q., and Opsteegh, T. (1995). Interannual and decadal variations of blocking activity in a quasi-geostrophic model. *Tellus* **47,** 941–954.

Livezey, R. E., and Chen, W. Y. (1983). Statistical field significance and its determination by Monte Carlo techniques. *Mon. Weather Rev.* **111,** 46–59.

Livezey, R. E., and Mo, K. C. (1987). Tropical–extratropical teleconnections during the northern hemisphere winter. Part II: Relationships between monthly mean northern hemisphere circulation patterns and proxies for tropical convection. *Mon. Weather Rev.* **115,** 3115–3132.

Lorenz, E. N. (1990). Can chaos and intransitivity lead to interannual variability? *Tellus* **42A,** 378–389.

Madden, R. A., Shea, D. J., Branstator, G. W., Tribbia, J. J., and Weber, R. O. (1993). The effects of imperfect spatial and temporal sampling on estimates of the global mean temperature: Experiments with model data. *J. Climate* **6,** 1057–1066.

Maier-Reimer, E., and Mikolajewicz, U. (1989). Experiments with an OGCM on the cause of the Younger-Dryas. *In* "Oceanography" (A. Ayala-Castanares, W. Wooster, and A. Yanez-Arancibia, eds.). UNAM Press, Mexico pp. 87–100.

Manabe, S., Stouffer, R. J., Spelman, M. J., and Bryan, K. (1991). Transient responses of a coupled ocean–atmosphere model to gradual changes of atmospheric CO_2. Part I: Annual mean response. *J. Climate* **4,** 785–818.

Mann, M. E., (1998). A Study of Ocean-Atmosphere Interaction and Low Frequency Variability of the Climate System. Ph.D. Thesis, Yale University, New Haven, CT.

Mann, M. E., and Lees, J. (1996). Robust estimation of background noise and signal detection in climatic time series. *Clim. Change* **33,** 409–445.

Mann, M. E., and Park, J. (1993). Spatial correlations of interdecadal variation in global surface temperatures. *Geophys. Res. Lett.* **20,** 1055–1058.

Mann, M. E., and Park, J. (1994). Global-scale modes of surface temperature variability on interannual to century timescales. *J. Geophys. Res.* **99,** 25819–25833.

Mann, M. E., and Park, J. (1996a). Greenhouse warming and changes in the seasonal cycle of temperature: Model versus observations. *Geophys. Res. Lett.* **23,** 1111–1114.

Mann, M. E., and Park, J. (1996b). Join spatio-temporal modes of surface temperature and sea level pressure variability in the Northern Hemisphere during the last century. *J. Climate* **9,** 2137–2162.

Mann, M. E., Lall, U., and Saltzman, B. (1995a). Decadal-to-century scale climate variability: Insights into the rise and fall of the Great Salt Lake. *Geophys. Res. Lett.* **22,** 937–940.

Mann, M. E., Park, J., and Bradley, R. (1995b). Global interdecadal and century-scale oscillations during the past five centuries. *Nature* **378,** 266–270.

Marple, S. L., Jr. (1987). "Digital Spectral Analysis with Applications." Prentice-Hall, Englewood Cliffs, NJ, 1987.

Marshall, S., Mann, M. E., Oglesby, R. J., and Saltzman, B. (1995). A comparison of the CCM1-simulated climates for pre-industrial and present-day CO_2 levels. *Glob. Planet. Change* **10,** 163–180.

Mehta, V. M., and Delworth, T. (1995). Decadal variability of the tropical Atlantic ocean surface temperature in shipboard measurements and in a global ocean–atmosphere model. *J. Climate* **8,** 172–190.

Mitra, K., Mukherji, S., and Dutta, S. N. (1991). Some indications of 18.6 year luni-solar and 10–11 year solar cycles in rainfall in northwest India, the plains of Uttar Pradesh and north-central India. *Internat. J. Climatol.* **11,** 645–652.

Moron, V., Vautard, V., and Ghil, M. (1997). Trends, interdecadal and interannual oscillations in global sea-surface temperatures. *Clim. Dyn.* **14,** 545–569.

Mysak, L. A., and Power, S. B. (1992). Sea-ice anomalies in the western Arctic and Greenland–Iceland Sea and their relation to an interdecadal climate cycle. *Climatol. Bull.* **26,** 147–176.

Mysak, L. A., Stocker, T. F., and Huang, F. (1993). Century-scale variability in a randomly forced, two-dimensional thermohaline ocean circulation model. *Clim. Dyn.* **8,** 103–116.

Namias, J. (1983). Short period climatic variations. *In* "Collected Works of J. Namias, 1975 Through 1982, Vol. III." University of California, San Diego.

Naujokat, B. (1996). An update of the observed quasibiennial oscillation of the stratospheric winds over the tropics. *J. Atmos. Sci.* **43,** 1873–1877.

Newman, M., and Sardeshmukh, P. D. (1995). A caveat concerning singular value decomposition. *J. Climate* **8,** 352–360.

Oglesby, R. J., and Saltzman, B. (1992). Equilibrium climate statistics of a general circulation model as a function of atmospheric carbon dioxide. Part I: Geographic distributions of primary variables. *J. Climate* **5,** 66–92.

Palmer, T. N., and Sun, A. (1985). A modeling and observational study of the relationship between sea surface temperature in the northwest Atlantic and the atmospheric general circulation. *Quart. J. Roy. Meteor. Soc.* **111,** 947–975.

Park, J. (1992). Envelope estimation for quasi-periodic geophysical signals in noise: A multitaper approach. *In* "Statistics in the Environmental and Earth Sciences" (A. T. Walden and P. Guttorp, eds.). Edward Arnold, London, pp. 189–219.

Park, J., and Maasch, K. A. (1993). Plio-Pleistocene time evolution of the 100-kyr cycle in marine paleoclimate records. *J. Geophys. Res.* **98,** 447–461.

Park, J., and Mann, M. E. (1999). Interannual temperature events and shifts in global temperature: A multiple wavelet correlation approach. *Earth Interactions*, in press.

Park, J., Lindberg, C. R., and Vernon III, F. L. (1987). Multitaper spectral analysis of high-frequency seismograms. *J. Geophys. Res.* **92,** 12675–12684.

Parker, D. E., Folland, C. F., and Jackson, M. (1995). Marine surface temperature: Observed variations and data requirements. *Clim. Change* **31,** 559–600.

Pedlosky, J. (1987). "Geophysical Fluid Dynamics." Springer-Verlag, New York.

Peixoto, J. P., and Oort, A. H. (1992). "Physics of Climate." American Institute of Physics, New York.

Penland, C. (1989). Random forcing and forecast-ing using Principal Oscillation Pattern Analysis. *Mon. Wea. Rev.* **117,** 2165–2185.

Percival, D. B., and Walden, A. T. (1993). "Spectral Analysis for Physical Applications." Cambridge Univ. Press, Cambridge, UK, 1993.

Philander, S. G. H. (1990). "El Niño, La Niña, and the Southern Oscillation." Academic Press, New York, 1990.

Pierce, D. W., Barnett, T. P., and Mikolajewicz, U. (1995). Competing roles of heat and freshwater flux in forcing thermohaline oscillations. *J. Phys. Oceanogr.* **25,** 2046–2064.

Preisendorfer, R. W. (1988). Principal component analysis in meteorology and oceanography. Development in Atmospheric Science **17**. Elsevier, Amsterdam.

Quinn, W. H., and Neal, V. T. (1992). The historical record of El Niño events. *In* "Climate Since A.D. 1500" (R. S. Bradley and P. D. Jones, eds.). Routledge, Boston, pp. 623–648.

Quon, C., and Ghil, M. (1995). Multiple equilibria and stable oscillations in thermosolutal convection at small aspect ratio. *J. Fluid Mech.* **291,** 33–56.

Rajagopalan, B., and Lall, U. (1995). Seasonality of precipitation along a meridian in the western United States. *Geophys. Res. Lett.* **22,** 1081–1084.

Rajagopalan, B., Cook, E., and Cane, M. A. (1996). Joint spatiotemporal modes of U.S. PDSI and Pacific SST variability. *Eos Suppl. AGU* **77,** 126.

Rajagopalan, B., Mann, M. E., and Lall, U. (1998). A multivariate frequency-domain approach to long lead climate forecasting. *Weather and Forecasting* **13,** 58–74.

Richman, M. B. (1986). Rotation of principal components. *J. Climatol.* **6,** 293–355.

Robock, A. (1996). Stratospheric control of climate. *Science* **272,** 972–973.

Roebber, P. J. (1995). Climate variability in a low-order coupled atmosphere–ocean model. *Tellus* **47,** 473–494.

Rogers, J. C. (1984). The association between the North Atlantic Oscillation and the Southern Oscillation in the Northern Hemisphere. *Mon. Weather Rev.* **112,** 1999–2015.

Ropelewski, C. F., and Halpert, M. S. (1987). Global and regional scale precipitation patterns associated with the El Niño–Southern Oscillation. *Mon. Weather Rev.* **115,** 1606–1626.

Ropelewski, C. F., Halpert, M. S., and Wang, X. (1992). Observed tropospheric biennial variability and its relationship to the Southern Oscillation. *J. Climate* **5,** 594–614.

Royer, T. C. (1993). High-latitude oceanic variability associated with the 18.6 year nodal tide. *J. Geophys. Res.* **98,** 4639–4644.

Saltzman, B. (1982). Stochastically-driven climatic fluctuations in the sea-ice, ocean temperature, CO_2 feedback system. *Tellus* **34,** 97–112.

Saltzman, B., and Moritz, R. E. (1980). A time-dependent climatic feedback system involving sea-ice extent, ocean temperature, and CO_2. *Tellus* **32,** 93–118.

Saltzman, B., Sutera, A., and Evenson, A. (1981). Structural stochastic stability of a simple auto-oscillatory climatic feedback system. *J. Atmospheric Sci.* **38,** 494–503.

Saravanan, R., and McWilliams, J. C. (1995). Multiple equilibria, natural variability, and climate transitions in an idealized ocean–atmosphere model. *J. Climate* **8,** 2296–2323.

Schlesinger, M. E., and Ramankutty, N. (1994). An oscillation in the global climate system of period 65–70 years. *Nature* **367,** 723–726.

Schmidt, G. A., and Mysak, L. A. (1996). The stability of a zonally averaged thermohaline circulation model. *Tellus* **48A,** 158–178.

Slowey, N. C., and Crowley, T. J. (1995). *Geophys. Res. Let.* **22,** 2345–2348.

Stocker, T. F. (1996). An overview of century time-scale variability in the climate system: Observations and models. *In* "Decadal Climate Variability: Dynamics and Predictability," NATO ASI, series 1 (D. L. T. Anderson and J. Willebrand, eds.). Springer-Verlag, New York.

Stocker, T. F., Wright, D. G., and Mysak, L. A. (1992). A zonally averaged coupled ocean–atmosphere model for paleoclimate studies. *J. Climate* **5,** 773–797.

Tanimoto, Y., Iwasaka, N., Hanawa, K., and Toba, Y. (1993). Characteristic variations of sea surface temperature with multiple time scales in the North Pacific. *J. Climate* **6,** 1153–1160.

Thomson, D. J. (1982). Spectrum estimation and harmonic analysis. *IEEE Proc.* **70,** 1055–1096.

Thomson, D. J. (1990). Time-series analysis of Holocene climate data. *Philos. Trans. Roy. Soc. London Ser. A.* **330,** 601–616.

Thomson, D. J. (1995). The seasons, global temperature, and precession. *Science* **268,** 59–68.

Thompson, L. (1992). *In* "Climate Since A.D. 1500" (R. S. Bradley and P. D. Jones, eds.). Routledge and Kegan Paul, Boston, pp. 517–548.

Thompson, R. (1995). Complex demodulation and the estimation of the changing continentality of Europe's climate. *Internat. J. Climatol.* **15,** 175.

Tinsley, B. A. (1988). The solar cycle and QBO influence on the latitude of storm tracks in the north Atlantic. *Geophys. Res. Lett.* **15,** 409.

Tourre, Y., Rajagopalan, B., and Kushnir, Y. (1999). Dominant patterns of climate variability in the Atlantic over the last 136 years. *J. Climate*, in press.
Trenberth, K. E. (1983). What are the seasons? *Bull. Am. Meteorol. Soc.* **64,** 1276–1282.
Trenberth, K. E. (1990). Recent observed interdecadal climate changes in the northern hemisphere. *Bull. Am. Meteorol. Soc.* **71,** 988–993.
Trenberth, K. E., and Hoar, T. J. (1995). The 1990–1995 El Niño-Southern Oscillation Event: Longest on record. *Geophys. Res. Lett.* **23,** 57–60.
Trenberth, K. E., and Hurrell, J. W. (1994). Decadal atmosphere–ocean variations in the Pacific. *Clim. Dyn.* **9,** 303–319.
Trenberth, K. E., and Paolino, D. A. (1980). The Northern Hemisphere sea-level pressure data set: Trends, errors and discontinuities. *Mon. Weather Rev.* **108,** 855–872.
Trenberth, K. E., and Shea, D. J. (1987). On the evolution of the Southern Oscillation. *Mon. Weather Rev.* **115,** 3078–3096.
Trenberth, K. E., and Shin, W.-T. K. (1984). Quasibiennial fluctuations in sea level pressures over the Northern Hemisphere. *Mon. Weather Rev.* **112,** 761–777.
Tziperman, E., Stone, L., Cane, M. A., and Jarsoh, H. (1994). El Niño chaos: Overlapping of resonances between the seasonal cycle and the Pacific ocean–atmosphere oscillator. *Science* **264,** 72–74.
Unal, Y. S., and Ghil, M. (1995). Interannual and interdecadal oscillation patterns in sea level. *Clim. Dyn.* **11,** 255–279.
Vautard, R., and Ghil, M. (1989). Singular spectrum analysis in nonlinear dynamics, with applications to paleoclimate time series. *Phys. D* **35,** 395–424.
Vautard, R., Yiou, P., and Ghil, M. (1992). Singular spectrum analysis: A toolkit for short noisy chaotic signals. *Phys. D* **58,** 95–126.
Venegas, S. A., Mysak, L. A., and Straub, D. N. (1996). Evidence for interannual and interdecadal climate variability in the South Atlantic. *Geophys. Res. Lett.* **23,** 2673–2676.
Vines, R. G. (1986). Rainfall patterns in India. *J. Climatol.* **6,** 135–138.
Von Storch, J. S. (1994). Interdecadal variability in a global coupled model. *Tellus* **46,** 419–432.
Von Storch, H., Burger, G., Schnur, R., and Von Storch, J. S. (1995). Principal Oscillation Patterns: A Review. *J. Climate* **8,** 377–399.
Wallace, J. M., and Dickinson, R. E. (1972). Empirical orthogonal representation of time series in the frequency domain. Part I: Theoretical considerations. *J. Appl. Meteorol.* **11,** 887–892.
Wallace, J. M., and Gutzler, D. S. (1981). Teleconnections in the geopotential height field during the northern hemisphere winter. *Mon. Weather Rev.* **109,** 784–812.
Wallace, J. M., Smith, C., and Bretherton, C. S. (1992). Singular value decomposition of wintertime sea surface temperatures and 500-mb height anomalies. *J. Climate* **5,** 561–576.
Weare, B. C., and Jasstrom, J. S. (1982). Examples of extended empirical orthogonal function analyses. *Mon. Weather Rev.* **110,** 481–485.
Weaver, A. J., and Sarachik, E. (1991). Evidence of decadel variability in an ocean general circulation model: An advective mechanism. *Atmos. Ocean* **29,** 197–231.
Weaver, A. J., Sarachik, E. S., and Marotzke, J. (1991). Freshwater flux forcing of decadal and interdecadal oceanic variability. *Nature* **353,** 836–838.
Wigley, T. L., and Raper, R. (1990). Natural variability of the climate system and detection of the greenhouse effect. *Nature* **244,** 324–327.
Wikle, C. K., and Cressie, N. (1996). A spatially descriptive, temporally dynamic statistical model with applications to atmospheric processes. Ph.D. dissertation, Iowa State University, Ames, IA.

Xu, J. S. (1993). The joint modes of the coupled atmosphere–ocean system observed from 1967–1986. *J. Climate* **6,** 816–838.

Yang, J., and Huang, R. X. (1996). Decadal oscillations driven by the annual cycle in a zonally-averaged coupled ocean–ice model. *Geophys. Rev. Lett.* **23,** 269–272.

Yang, J., and Neelin, J. D. (1993). Sea-ice interaction with the thermohaline circulation. *Geophys. Res. Lett.* **20,** 217–220.

Yiou, P., Genthon, C., Jouzel, J., Ghil, M., Le Treut, H., Barnola, J.M., Lorius, C., and Korotkevitch, Y. N. (1991). High-frequency paleovariability in climate and in CO_2 levels from Vostok ice-core records. *J. Geophys. Res.* **96B,** 20365–20378.

Yiou, P., Ghil, M., Jouzel, J., Paillard, D., and Vautard, R. (1994). Nonlinear variability of the climatic system, from singular and power spectra of Late Quaternary records. *Clim. Dyn.* **9,** 371–389.

Zhang, S., Lin, C. A., and Greatbatch, R. (1995). A decadal oscillation due to the coupling between an ocean circulation model and a thermodynamic sea-ice model. *J. Marine Res.* **53,** 79–106.

NUMERICAL MODELS OF CRUSTAL DEFORMATION IN SEISMIC ZONES

STEVEN C. COHEN

Geodynamics Branch
Goddard Space Flight Center
Greenbelt, Maryland 20771

1. INTRODUCTION

During the past few decades remarkable advances have been made in measuring and modeling crustal deformation. The accuracy of geodetic observations now approaches millimeters over spatial scales in excess of 1000 km and continuous monitoring of surface motion using space techniques is replacing infrequent ground-based resurveying. Computational models of the crustal deformations in seismically active zones, particularly at tectonic plate boundaries, have become standard research tools that are used to infer such seismotectonic parameters as the amplitude and depth distribution of fault slip, subsurface fault geometry, and the rheological parameters of the lower crust and upper mantle. The tools that are available range from analytical solutions for the displacements due to uniform coseismic fault slip in an elastic half-space to finite-element solutions for the deformation throughout the earthquake cycle in a rheological nonlinear, spatially heterogenous or spherically layered Earth. This chapter is a tutorial on the physical and mathematical ideas and techniques that are used in modeling crustal deformation and a critical review of both general and locale-specific applications.

The earliest models of seismic-zone crustal deformation dealt with static crustal displacements that accompany earthquakes. They owe their intellectual heritage to Reid's elastic rebound theory of earthquakes (Reid, 1911), the dislocation theory concepts of solid-state physics, and the mathematical apparatus of Green's functions. From the outset, an objective in developing earthquake models was to provide a mathematical description of how displacement and other deformation parameters, such as strain and tilt, vary with location and time. As we shall see, these spatial and temporal variations depend on the earthquake slip vector, fault geometry, and the mechanical properties of the crust and mantle. Since seismological and geological, as well as geodetic, observations provide measurements of the deformation, empirical observations can be used in conjunction with theoretical models to derive estimates of the parameters which describe an earthquake. These parameters can, in turn, be used in kinematic and dynamic models of strain accumulation and release. Beyond their direct application to studies of seismotectonic processes, models of crustal and subcrustal deformation are also useful for estimating the effects of earthquakes on other geophysical phenomena such as the change in the Earth's moment of inertia and spin axis.

Efforts to understand earthquake-related crustal deformations have now extended far beyond the computation of coseismic movements. Much of the theoretical effort of the past few decades has focused on modeling the deformations that occur between major seismic events. The term "earthquake cycle" is often used to refer to the repeated accumulation and release of stress and strain. Of course, earthquakes are not periodic and successive events are not repeats of one another, even though it is sometimes useful to assume periodicity in mathematical models. However, certain features are common characteristics of major earthquake cycles. Immediately after an earthquake, the crustal movement may be rapid compared to the preseismic or average interseismic motion. In the near-field, the period of rapid postseismic deformation is usually a few decades or less, often much less, whereas the interseismic period, i.e., the time between major earthquakes, is usually centuries. Commonly, the deformation progresses smoothly from rapid postseismic movement to more gradual interseismic movement, though the postseismic deformation may occur over several time scales. Occasionally, a period of rapid preseismic deformation, lasting from hours to months, precedes the earthquake occurrence.

Crustal and subcrustal deformation depends not only on the fault parameters, but also on the rheological properties of the Earth's crust and mantle, possibly to depths of a few hundred kilometers. In the simplest models, the Earth is taken to be a homogeneous or layered elastic body, the deformations are driven only by the motion of the tectonic plates, and strain accumulates at a constant rate between earthquakes. This model explains many of the first-order features of earthquakes at tectonic plate boundaries and is the essential aspect of the well-known elastic rebound theory, at least in its most commonly expressed form. However, rock mechanics, plate tectonics, and postglacial rebound studies show that anelastic flow will occur in the upper mantle and the lower crust. The anelastic flow can be represented by viscoelastic or even plastic rheological laws that give rise to a time-dependent response to faulting even when there is no further slip on the fault after the earthquake. Time-dependent aseismic fault slip, possibly at depths below the seismogenic layer, provides another mechanism for time-dependent deformation. Consequently, the study of the spatial-temporal pattern of crustal deformation, particularly that occurring postseismically when the signal is large, can provide insight into the rheological properties of the crust and uppermost mantle.

Models of postseismic and interseismic deformation are more difficult to formulate and test than those for coseismic deformations. Strain accumulates in response to tectonic plate motion and is affected by the rheological response of the Earth to loading and by slip on nearby faults or fault segments. Gravity may play an important role in controlling vertical motion, particularly over long times. The boundary conditions which act on the system are not entirely obvious, particularly at convergent plate boundaries where the influence of the subducting slab may be considerable. Controversies exists concerning the degree to which permanent deformation accumulates over repeated earthquake cycles, and only modest progress has been made on incorporating erosional, depositional, and accretionary processes into earthquake models.

The road map for this article follows. At the outset, we will examine the features of a simple model of the surface deformation that accompanies faulting on an infinitely long, vertical, strike-slip fault embedded in an elastic half-space. This model incorporates many of the ideas which are embedded in more sophisticated models and finds practical application in obtaining first-order estimates of faulting parameters. Following this introduction, we will undertake a more systematic discussion of modeling theory and results. A key physical concept is the representation of fault slip as a solid-state "dislocation" and a key mathematical feature is the use of Green's functions for solving differential equations. In some cases, the Green's functions can be used to derive closed-form solutions to the equilibrium equations. These analytical solutions are useful not only because they provide closed-form (or quadrature) formulas for the displacements, but also because they help in developing physical insight into the deformation process and because they provide an avenue for developing solutions to more complex problems. The obvious limitation of the analytical approach is that it requires that the geometry and boundary conditions be simple. More realistic situations often require the introduction of numerical approaches, notably the finite-element method. The strength of the finite-element method is that it readily accommodates complex boundary conditions, structural heterogeneities, rheological variations and nonlinearities, and other complexities. For dealing with displacements over large distances, spherical, rather than rectangular, Earth models are required to take into account the planet's curvature and finite size.

After developing models for coseismic deformation, we will consider postseismic rebound. We will pay particular attention to two major "deep" mechanisms of postseismic deformation, namely, viscoelastic flow at depth and fault creep below the seismically locked layer. We will see that determining what mechanism is responsible for observed postseismic deformation is an important, often challenging aspect of crustal deformation research. Since models of postseismic rebound necessarily deal with time-dependent processes, they are a natural springboard to considering the entire earthquake cycle. We will review how features of crustal deformation vary through the cycle, how models differ from one another in important details, and some of the questions that are active research topics.

After dealing with crustal deformation in a general sense, we will turn to some applications of the theoretical techniques to specific environments. Here we will see examples of how new technologies such as the Global Positioning System (GPS) and Satellite Radar Interferometry provide high accuracy and temporally rapid and spatially dense data. However, data on the entire earthquake cycle are still limited by the long recurrence intervals between major seismic events; therefore, historical data remain a valuable asset. Older position data, even when noisy, become increasingly usable as time passes and the cumulative deformation increases to a point where it is above the noise level in the observations.

My intent is to present an introduction and review of numerical modeling that is comprehensible to new workers in the field and yet useful to seasoned researchers who have had only limited exposure to the type of modeling and data interpretation that we will be considering. Accordingly, the arguments I use in developing equations will be physically based where possible and I will, at times, eschew mathematical rigor. However, I hope to provide a sufficiently strong introduction to the physics and mathematics of numerical modeling that the reader will be able to apply existing numerical models to research problems of current importance or develop his or her own new models. I will assume that the reader is conversant with basic aspects of physical and structural geology, seismology, linear elasticity and viscoelasticity, vector and matrix mathematics, differential and integral calculus, and Laplace and Fourier transform techniques. Most of the models we will review are kinematic in the sense that slip is imposed on a fault either to represent actual fault slip or as a mathematical artifact to represent interseismic loading. The calculations are aimed at determining the temporal and spatial pattern of deformation away from the fault. The most important exception, discussed in Section 3.4.4, involves the use of a constitutive law to represent fault friction with model calculations predicting fault slip as well as off-fault deformation. A notational issue should be mentioned at the outset. The reader will probably note that sometimes the same physical quantity is represented by different symbols; for example, both u and V are used for displacement. At other times, different physical quantities are represented by the same symbol; for example, τ can represent either a relaxation time or stress. This is done to maintain consistency with the cited research papers. The context makes the usage clear.

1.1. Elastic Half-Space Model of the Earthquake Cycle for an Infinitely Long Strike-Slip Fault: An Illustrative Model

To motivate the discussion in subsequent sections, we consider a simple example of fault-zone deformation, discussing several important results but deferring detailed derivations until later. Figure 1 shows an infinitely long, vertical, strike-slip fault in a uniform elastic half-space. The fault is locked between earthquakes at depths extending from the surface to the depth $z = -D$, below which the adjacent blocks are free to slip past one another. At great distances from the fault, the right-hand block moves into the paper at a constant relative velocity, $v/2$, while the left-hand block moves out of the paper with a constant relative velocity, $-v/2$. Earthquake slip is imposed at an evenly spaced time interval, T. The fault slip during an earthquake is $\Delta V_0 = vT$. Starting from a reference state immediately after an earthquake, the surface displacement, V, in the y direction (into the paper), varies with distance from fault, x, as

$$V(x,t) = \frac{vt}{\pi} \tan^{-1}\left(\frac{x}{D}\right), \qquad x > 0. \tag{1}$$

The only nonvanishing component of surface shear strain is

$$\varepsilon_{xy}(x,t) = \frac{1}{2}\frac{\partial V}{\partial x} = \frac{vt}{2\pi D}\frac{1}{1 + (x/D)^2}, \qquad x > 0. \tag{2}$$

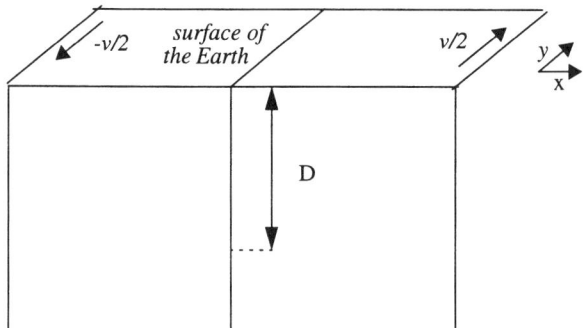

FIG. 1. Simple model for strike-slip faulting in an elastic half-space. The fault is loaded by the relative plate motion, v. Between earthquakes, the fault is locked from the surface to depth, D, but is free to slip below D.

The coseismic elastic displacement (for $x > 0$) is the complement of (1) in the sense that $V(t = 0) + V(t = T) = vT/2$, i.e.,

$$V(t = 0) = \Delta V_0 \left[\frac{1}{2} - \frac{1}{\pi} \tan^{-1}\left(\frac{x}{D}\right) \right]. \tag{3}$$

Figure 2 shows how the coseismic deformation varies with distance from the fault. There are several important implications of (1)–(3), specifically.

(1) The interseismic surface displacement and strain increase at a constant rate.

(2) The peak in the shear strain rate occurs at the fault trace and has the value $v/(2\pi D)$; strain and strain rate decrease to half their peak value at $x = D$. Similarly, the coseismic displacement drops to half its peak value at $x = D$.

(3) For $x \gg D$, the coseismic displacement depends on the inverse first power of distance from the fault, whereas the strain decreases with the inverse second power of distance. However, this conclusion is based on the assumption that the fault is infinitely long; therefore, all observation points are essentially in the near-field. For finite sources, the displacements and strains decrease more rapidly with distance.

Some typical numerical values for the parameters for a large earthquake are $v = 30$ mm/yr, $T = 200$ yr (hence, $\Delta V_0 = 6$ m), and $D = 10$ km. The coseismic interseismic near-field strain rate is $\dot{\varepsilon}_{xy}(0, t) \cong 5 \times 10^{-7}$ yr^{-1}. The total strain accumulated between earthquakes is 1.0×10^{-4} ($= 100$ μstrain).

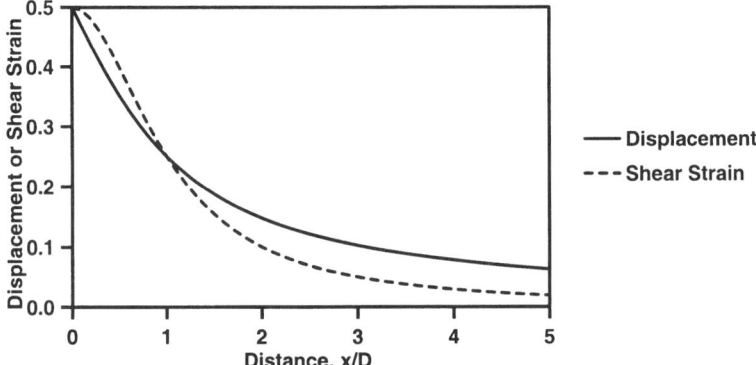

FIG. 2. Coseismic displacement, $V/\Delta V_0$, and absolute value of shear strain, $[|\varepsilon|/(\Delta V_0/\pi D)]$, versus distance, x/D, from the fault for the model shown in Figure 1.

There are many ways in which the real Earth may depart from the simple model we have presented here. Some of the more important of these departures include:

(1) The interseismic deformation rate might not be constant and earthquakes are not periodic.

(2) The spatial distribution of deformation can be more complex than the simple model predicts. For example, variations in the fault geometry and the fault-slip magnitude and direction can complicate the deformation pattern. The deformation pattern also varies at the edges of finite-length faults where there is a transition from slip to no-slip conditions. Different portions of a fault can rupture at different times, giving rise to an interaction between fault segments, and there might be an interaction with other nearby faults as well.

(3) Other stresses may interact with the tectonic ones. Most notable is the role played by gravity in modulating the long-term vertical deformation between earthquakes.

2. Coseismic Deformation

2.1. Elastic Dislocation Theory

In this section, we consider an analytical approach to modeling crustal deformation; later, we will turn to numerical modeling techniques. The starting point is a calculation of the quasi-static displacements that occur in an earthquake. We take advantage of the facts (1) that there is a well-known solution for the displacements due to a point force acting in an infinite homogeneous elastic body, (2) that fault slip or displacement may be regarded as a dislocation, or plane of displacement discontinuity, in a solid, and (3) that the dislocation is dynamically equivalent to a double-couple force system. To calculate the coseismic deformation, we proceed as follows:

(1) First we compute the displacement due to a force acting on a single point in a uniform elastic space. The governing differential equation is the equation of motion (or equilibrium) for a continuum system. This equation can be solved by Green's functions techniques.

(2) Having the solution for a single point force, it is a straightforward matter to compute the displacement due to both a force couple and, subsequently, a pair of orthogonal force couples. Force couples are needed to avoid having an unbalanced moment and torque associated with a single force. As in standard seismology problems, the double couple results in the

observed ambiguity between the fault plane and the orthogonal auxiliary plane (see, e.g., Lay and Wallace, 1995). Since the dislocation is not confined to a point, but rather occurs over a finite fault plane, the displacement pattern is obtained by integrating this solution for a point source over the entire fault plane. This results in a formula in which the displacement is expressed as an integral over the fault plane of the dislocation times a weighting function which is related to stress.

(3) The real problem to be considered is not deformation in a whole-space but rather deformation in an elastic half-space, where the surface of the Earth is the half-space boundary. The half-space problem can be solved by adding an imaginary "image source" (Weertman, 1964) to the whole Earth problem and then adding the effects of a surface normal force. The image source solution is readily obtained from the solution to the whole-space problem simply by replacing the source position by an image position as described below. The sum of the original solution plus the image solution satisfies the boundary condition that traction vanish on the Earth's surface, but this composite has twice the vertical normal stress at the surface as the original solution. Another solution to the differential equation must be added to the combined solution to cancel this normal stress. This is obtained through the use of the Galerkin vector, defined below.

As we will see, the solution can be expressed in terms of certain surface integrals which can be evaluated analytically in some cases of practical interest, or numerically in others. Once the elastic half-space problem is solved, it will serve later as the departure point for developing models with more complex geometries and rheologies.

Now to fill in some details. The first task is to write down the equation for the displacement due to a point force. The governing equation of motion is

$$\rho \frac{\partial^2 u_i}{\partial t^2} = f_i + \frac{\partial \tau_{ij}}{\partial x_j}, \qquad (4)$$

where ρ is the density, u_i, f_i, and x_i are the ith components of displacement, body force per unit volume, and position, respectively, and t is time. The repeated index implies a summation. The stress, τ_{ij}, and strain, $\varepsilon_{ij} = \frac{1}{2}(\partial u_i/\partial x_j + \partial u_j/\partial x_i)$, are related by

$$\tau_{ij} = c_{ijpq} \varepsilon_{pq}, \qquad (5)$$

where, for an isotropic medium,

$$c_{ijpq} = \lambda \delta_{ij} \delta_{pq} + \mu(\delta_{ip}\delta_{jq} + \delta_{iq}\delta_{jp}), \qquad (6)$$

and where λ and μ are the Lamé elastic constants and δ is the Kronecker delta. It follows that

$$\rho\frac{\partial^2 u_i}{\partial t^2} = f_i + (\lambda + \mu)\frac{\partial^2 u_j}{\partial x_j \partial x_i} + \mu\frac{\partial^2 u_i}{\partial x_j \partial x_j}. \tag{7}$$

Techniques for solving (7), notably the Green's functions formulation, are discussed in standard seismological texts (e.g., Aki and Richards, 1980). Since we are not concerned with seismic wave propagation, we can set the term on the left-hand side of the equation to zero. For a point force, F, at position ξ_1, ξ_2, ξ_3 compensated by distributed forces at infinity so the system is in static equilibrium, the Somigliana tensor solution is (Steketee, 1958a, b; Kasahara, 1981; Maruyama, 1973):

$$u_i^k = \frac{F_k}{4\pi\mu}\left[\delta_{ik}\frac{1}{r} - \frac{\lambda + \mu}{2(\lambda + 2\mu)}\frac{\partial^2 r}{\partial x_i \partial x_k}\right], \tag{8}$$

where $r = [(x_1 - \xi_1)^2 + (x_2 - \xi_2)^2 + (x_3 - \xi_3)^2]^{1/2}$ and the indices i and k indicate the direction of the displacement and applied force, respectively.

Now that we have the displacement due to a point force, we have to consider how to relate this to faulting. To answer this question, we appeal to the intuitive discussion of Maruyama (1973) and Kashahara (1981), who considered how the displacement must vary with distance near a fault. The situation is shown in Figure 3a. Faulting is represented by the displacement discontinuity or dislocation, i.e., the dashed line in the figure. The solid line represents an approximation to the actual situation which approaches the true curve in the limit that Δ approaches zero. We take the slope of the solid line to calculate the shear strain, $\varepsilon_{21} = \varepsilon_{12}$, then from the elasticity relation, (5), calculate the shear stress. The result is shown diagrammatically in Figure 3(b). The shear stress is then used in the force equilibrium equations, for example,

$$f_1 + \frac{\partial \tau_{11}}{\partial x_1} + \frac{\partial \tau_{12}}{\partial x_2} + \frac{\partial \tau_{13}}{\partial x_3} = 0. \tag{9}$$

Ignoring shear stress terms other than that involving $\tau_{21} = \tau_{21}$, we get

$$f_1 = \frac{\partial \tau_{12}}{\partial x_2} = -\mu\frac{\partial^2 u_1}{\partial x_2^2}, \tag{10}$$

which is shown graphically in Figure 3c. As the solid curve at the top of the figure approaches the dashed curve, the peaks in the force system become

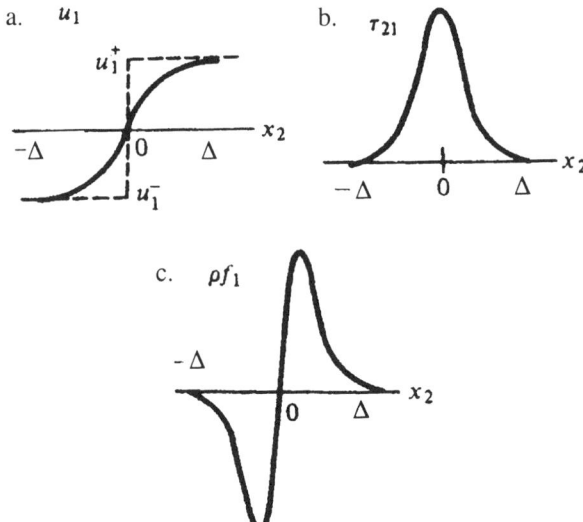

FIG. 3. Schematic view of the surface of a shear dislocation. (a) The displacement is shown as a solid line which approaches the limit shown as the dashed line as $\Delta \to 0$. (b) The resulting shear stress. (c) Force. The displacement and coordinates components are designated by subscripts, cf. Figures 1 and 2. (Kashahara, 1981; reproduced with permission from "Earthquake Mechanics," ©1981 Cambridge University Press.)

higher and thinner, producing a force couple. A similar couple occurs in the perpendicular direction. In the limit that $\Delta \Rightarrow 0$, the force system becomes a double couple.

It follows from the preceding argument that in order to calculate the displacement field due to faulting, we must consider the displacement due to force couples. The displacement due to two equal magnitude but oppositely directed point forces located at positions $\xi_1 - h$ and $\xi_1 + h$ is

$$u_i(\xi_1 + h) - u_i(\xi_1 - h) \cong 2h \frac{\partial u_i}{\partial \xi_1}, \tag{11}$$

where the u's are given by (8). For the double couple, the resulting displacement is proportional to $\partial u_i^2/\partial \xi_1 + \partial u_i^1/\partial \xi_2$, where the superscripts refer to the two components of the double-couple system. A key point is that the displacement due to the couple is proportional to the gradient of the displacement due to a single force. The displacement gradient or strain is proportional to the stress. The preceding argument has been made more rigorous (see, e.g., Eshelby, 1973). For the general case in which a set of dislocation sources is distributed over a finite dislocation surface, Σ, with position-dependent dislocation defined by $\Delta u = u^+ - u^-$, the resulting displacement field is a convolution of the dislocation function and the stress (Kasahara, 1981), i.e.,

$$u_m(X) = \iint \Delta u_k(\xi) \tau_{kl}^m(X, \xi) v_l \, d\Sigma, \tag{12}$$

where τ_{kl} is stress. From (8),

$$\tau_{kl} = \frac{1}{4\pi}\left[\frac{\mu}{\lambda + 2\mu}\left(-\delta_{kl}\frac{r_m}{r^3} + \delta_{mk}\frac{r_l}{r^3} + \delta_{ml}\frac{r_k}{r^3}\right) + 3\left(\frac{\lambda + \mu}{\lambda + 2\mu}\right)\frac{r_k r_l r_m}{r^5}\right], \tag{13}$$

X is the three-dimensional location of an observation point, $r_m = x_m - \xi_m$ (with similar definitions for r_l and r_k), and v_l is the direction cosine of the fault plane. Those familiar with potential theory will note the similarity between (12) and the Green's functions formulation of potential problems. Equation (12) is known as Volterra's formula. An implication of (12) and (13) is that the displacement decreases as the inverse square of distance at large distances from the source. Compare this result, for a finite-size source, with the more gradual decrease in displacement with distance for an infinitely long fault as given in Section 1.1. More details on the applications of Green's functions to both static and dynamic dislocations can be found in the textbook by Aki and Richards (1980). The reader who wishes to fill in the details in going from (8) to (13) should pay close attention to the indices when calculating the partial derivatives, particularly in cases such as $k = l$.

We now have an expression for the displacement due to a finite dislocation in an infinite elastic space. The next step is to consider the half-space case with the upper boundary of the half-space being the surface of the Earth. We already know that our preceding solution satisfies the governing differential equation. The problem is that it does not necessarily satisfy the surface boundary conditions that the traction and normal stress vanish at the Earth's surface. However, if we add to the preceding equation an identical equation which uses a source at an "image" location which is the mirror location of the real source, i.e., is at $(\xi_1, \xi_2, -\xi_3)$, then the shear stress at the Earth's surface vanishes. This superposition also results in a doubling of the normal stress at the surface, so unless the normal stress vanishes, the remaining task is to find another solution which exactly cancels this normal stress and adds no further shear. This latter problem is called the Boussinesq problem. Steketee (1958a) considers the Galerkin vector, Γ, related to displacement by

$$u_i = \frac{\partial^2 \Gamma_i}{\partial x_k \partial x_k} - \left(\frac{\lambda + \mu}{\lambda + 2\mu}\right) \frac{\partial^2 \Gamma_k}{\partial x_k \partial x_i}. \quad (14)$$

Each component of the Galerkin vector satisfies the biharmonic equation,

$$\frac{\partial \Gamma_i^4}{\partial x_j \partial x_j \partial x_k \partial x_k} = 0, \quad (15)$$

as can be verified by direct substitution. Because we are trying to add a solution which will cancel the normal stress at the surface of the Earth from the original and image source, we consider the vertical component Γ_3 of (15). A solution may be obtained by introducing the Fourier transform, $\bar{\Gamma}_3$, where,

$$\Gamma_3(x_1, x_2, x_3) = \frac{1}{2\pi} \iint_{-\infty}^{+\infty} \bar{\Gamma}_3(s_1, s_2, x_3) \exp[i(s_1 x_1 + s_2 x_2)] \, ds_1 \, ds_2 \quad (16a)$$

and

$$\bar{\Gamma}_3(s_1, s_2, x_3) = \frac{1}{2\pi} \iint_{-\infty}^{+\infty} \Gamma_3(x_1, x_2, x_3) \exp[-i(s_1 x_1 + s_2 x_2)] \, dx_1 \, dx_2. \quad (16b)$$

The differential equation for $\bar{\Gamma}_3$ is

$$\left[\frac{d^2}{dx_3^2} - (s_1^2 + s_2^2)\right]^2 \bar{\Gamma}_3 = 0. \quad (17)$$

The solution to (17) which is bound at large distances and which gives a vanishing shear stress at the surface is

$$\bar{\Gamma}_3 = B\left(2 - \left(\frac{\lambda + 2\mu}{\lambda + \mu}\right) + \left(\sqrt{s_1^2 + s_2^2}\right)x_3\right)\exp\left(-\sqrt{s_1^2 + s_2^2}\,x_3\right). \quad (18)$$

The coefficient, B, is chosen to make the normal stress for the total solution vanish at the surface. The remaining steps in the calculation are to invert the Fourier transform and do the required differentiations to get the displacement. The reader is referred to the paper by Steketee (1958a), where the details are presented; a few specific results will be presented in the next section. While the work of Steketee (1958a, b) is most often cited as the pioneering study in the application of dislocation theory to earthquake problems, there are other early studies which merit recognition. Examples include the work of Kasahara (1957), who derived the displacement field from an assumed stress distribution, and that of Rongved and Frasier (1958). A further discussion of the use of the method of images can be found in a paper by Mahrer (1984).

The elastic dislocation theory that we have begun to consider in this section is based, in a large part, on work in solid-state physics where a dislocation is a line of defects in the crystal lattice. Some of the solid-state physics terminology has been adopted by the geophysical community. An important item for discussing a dislocation is the Burgers vector, **b**, defined as follows (Kittel, 1971):

(1) Draw a curve with the solid body. The curve may be either closed or open, with the ends terminating on the surface of the body.

(2) Make a cut along a simple surface bounded by the line.

(3) Displace the material on one side of the surface relative to the other side by a direction and amount given by **b**.

(4) In regions where **b** is not parallel to the cut surface, there will be either a gap or an overlap between the two sides. Fill in material to remove the gap or remove material to eliminate the overlap, rejoin the two sides and while maintaining **b**, otherwise allow the material to return to equilibrium.

If the Burgers vector is oriented perpendicular to the line of the dislocation, then the dislocation is termed an edge dislocation; if it lies parallel, it is a screw dislocation. Strike-slip faults can be modeled as one or more screw dislocations, whereas reverse and normal faults can be modeled by a combination of edge dislocations.

2.2. Early Applications of Dislocation Theory

One of the first applications of the dislocation theory to field observations can be found in Chinnery (1961), where displacements are computed for strike-slip faulting on a rectangular vertical surface. Taking the fault to extend vertically (along the y_3 axis) from an upper depth, d, to a lower depth, D, and along the horizontal axis (y_1 axis), from $-L$ to $+L$, the displacements, u_k, are

$$\frac{u_1}{u} = \frac{y_2 t \alpha}{4\pi} \left\{ \frac{1}{s_1(s_1+q)} + \frac{(1+c)s_2 + (1-b)p + q}{s_2(s_2+p)^2} \right.$$
$$\left. - \frac{(p^2-q^2)(2s_2+p)}{2s_2^3(s_2+p)^2} \right\} - \frac{1}{4\pi} \qquad (19)$$
$$\cdot \left\{ \tan^{-1}\left(\frac{y_2 s_1}{qt}\right) + \tan^{-1}\frac{y_2 s_2}{pt} \right\} \bigg\|,$$

$$\frac{u_2}{u} = \frac{1}{4\pi}\{(1-\alpha)\ln(s_1+q) + (1-\alpha-\alpha c)\ln(s_2+p)\}$$
$$- \frac{\alpha}{4\pi}\left\{ \frac{(b+c)p-q}{(s_2+p)} + \frac{(p^2-q^2)}{2s_2(s_2+p)} \right\}$$
$$- \frac{y_2^2 \alpha}{4\pi}\left\{ \frac{1}{s_1(s_1+q)} + \frac{(1+c)s_2 + (1-b)p + q}{s_2(s_2+p)^2} \right. \qquad (20)$$
$$\left. - \frac{(p^2-q^2)(2s_2+p)}{2s_2^3(s_2+p)^2} \right\}\bigg\|,$$

and

$$\frac{u_3}{u} = \frac{y_2}{4\pi}\left\{ \alpha\left(\frac{1}{s_1}-\frac{1}{s_2}\right) + \frac{s_2(1+b) + 2\alpha p + 2q(1-\alpha)}{s_2(s_2+p)} \right.$$
$$\left. + \alpha\frac{(p^2-q^2)}{2s_2^3} \right\}\bigg\|, \qquad (21)$$

where the notation is close to that in Chinnery's paper. The fault slip is u, the observation point is at (y_1, y_2, y_3), and a source point has coordinates (x_1, x_2, x_3). Some auxiliary definitions are: $t = x_1 - y_1$, $p = x_3 + y_3$, $q = x_3 - y_3$, $s_1^2 = t^2 + y_2^2 + q^2$, $s_2^2 = t^2 + y_2^2 + p^2$, $\alpha = (\lambda+\mu)/(\lambda+2\mu)$, $b = (\lambda-\mu)/(\lambda+\mu)$, and $c = \mu\lambda/(\lambda+\mu)^2$. The symbol $\|$ indicates that the corresponding function should be evaluated at the limits of the double integration over the fault plane, i.e.,

$$f(x_1, x_3)\| = f(+L, D) - f(+L, d) - f(-L, D) + f(-L, d).$$

An example of the surface displacements due to a finite-length fault as computed using the preceding equations is shown in Figure 4a. Note the

a.
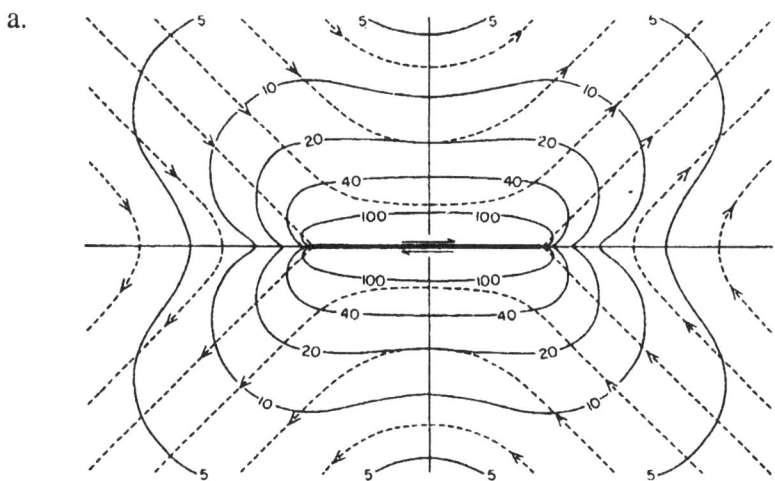

b.
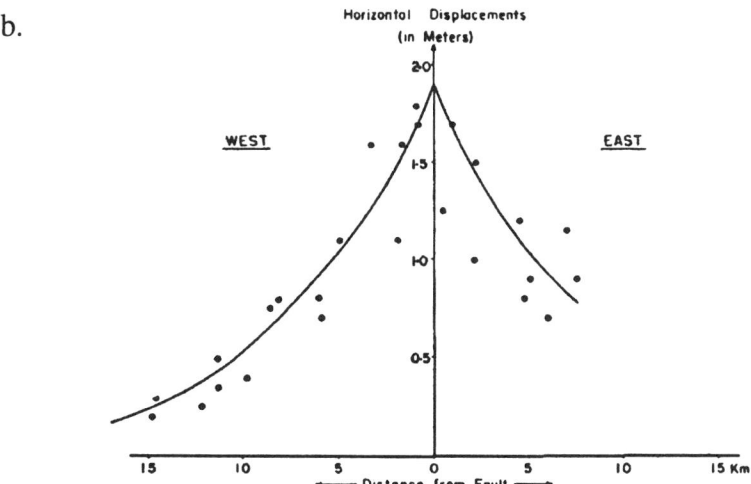

FIG. 4. (a) Horizontal displacements due to strike-slip faulting for $D/L = 0.1$. The solid lines show contours of displacement magnitude (in units of $10^{-3} \times$ slip) while the dashed lines show the direction of motion. (b) Observed and computed displacements for 1930 North Idu, Japan earthquake. For the theoretical calculations, $D = L = 12$ km. (Chinnery, 1961; reproduced with permission from *Bull. Seismol. Soc. Am.*, ©1961 Seismological Society of America.)

symmetries in the displacement pattern. Chinnery (1961) applied this model to earthquakes in Japan and San Francisco and derived estimates of the depth of faulting. Figure 4b shows the variation in horizontal coseismic displacement with distance from the fault for an earthquake that occurred along the Tanna fault, North Idu, Japan in 1930. For the theoretical calculations, both the depth of faulting and the fault half-length were taken to be 12 km. In a later paper, Chinnery (1963) computed the stress drop in a strike-slip earthquake by calculating the appropriate displacement gradients and applying the elastic constitutive equations to relate the strain change to stress drop (see also Sato and Matsu'ura, 1974). Barnett and Freund (1975) estimated a fault depth of about 8 km for the San Andreas fault by fitting a dislocation model to observed surface deformation. A minor defect in the physics of some of the early models is the occurrence of a stress singularity at the edge of the fault-slip plane due to the sudden transition from uniform slip to no slip. Sato (1972) avoided this problem by replacing the uniform-slip function by one which passes smoothly to zero at the ends of the fault, and Barnett and Freund (1975) avoided the singularity by introducing an integral constraint on the fault stress (see also Rice, 1968).

The solutions we have presented up to now are based on an elastic half-space model. Turcotte and Spence (1974) derived an alternative model based on a solution to the equilibrium equations within a thin plate. A deduction from this model is that strain accumulation is concentrated within about a fault-length distance from the source; however, the half-space models indicate otherwise—that the deformation is concentrated within a fault-depth distance. As pointed out by Savage (1975), the thin-plate approximation is not justified. The bottom of the lithospheric plate is bound to the asthenosphere, which is elastic in the short term following an earthquake, so thin-plate behavior is not expected.

The calculation of the deformation due to dip-slip faulting on a finite-length fault is more complex than the corresponding problem for strike-slip faulting and leads to fewer simple end-member formulas. Press (1965) derived equations for the far-field displacements resulting from various slip configurations by using Galerkin vectors previously obtained by Mindlin and Cheng (1950). He applied the model to several moderate to large earthquakes including the great 1964 Prince William Sound, Alaska earthquake. Although this model for that earthquake is not correct (he assumed a steeply dipping fault, but the earthquake occurred on the shallow dipping North America–Pacific Plate megathrust), his prediction that detectable displacements will occur at great distances from the source was an important impetus for other geodetic and seismological studies. It also spurred Mansinha and Smylie (1967), Mansinha *et al.*, (1979), and subsequent researchers to consider whether seismic events could excite the Chandler wobble. Some researchers have found temporal correlations between changes in polar motion and the occurrence of earthquakes, but the computed pole shifts have been too small to explain the observations; thus, most investigators now believe that earthquakes are not the major driver of the Chandler wobble.

Savage and Hastie (1966) integrated the point dislocation functions of Maruyama (1964) over a rectangular fault plane to derive the surface vertical displacements. They applied their model to the aforementioned Prince William Sound earthquake as well as to the 1954 Fairview Peak, Nevada, Montana ($M = 7.1$) and 1959 Hebgen Lake, Montana ($M = 7.1$) events. By fitting the predicted coseismic uplift in Alaska to data published by Plafker (1965), they derived the following parameters for the Prince William Sound earthquake: fault length, $2L = 600$ km, fault width, $W = 200$ km, depth to top of fault = 20 km, coseismic slip = 10 m, and dip angle = 9°. Although more recent analyses have refined these parameter determinations somewhat, the Savage and Hastie (1966) results marked a significant advance in deriving fault parameters for dipping faults from geodetic observations.

An important contribution to dislocation analysis was provided by Mansinha and Smylie (1971), who provided closed-form solutions to the general three-dimensional problem of slip on an inclined plane. The Mansinha and Smylie (1971) equations can be found in the original reference and in Savage (1980). They have been widely used despite having a few minor computational instabilities for certain geometries. The mathematical problems can be overcome by perturbing the input parameters by physically insignificant amounts. The published solutions are for displacements, but tilts and strains can be obtained by analytical or numerical differentiation. Iwasaki and Sato (1979) published explicit expressions for strain, but more general results can be found in Okada (1985, 1992). The latter papers present what are now the most frequently cited numerical modeling equations. They include expressions for displacements, strains, and tilts. They are numerically more stable than the earlier results and relax the assumption that the Lamé constants are equal. The equations for the displacement due to a point source are given in the Appendix. These and the corresponding equations for a finite-length fault can also be found in the original source (Okada, 1992). This reference also provides tables with explicit expressions for the displacement gradients that are needed to calculate strains and tilts. Both Okada (1985) and Okada (1992) provide some informative summaries of earlier developments in dislocation modeling. However, the tables of published analytical solutions that are presented there are incomplete. Notably absent are the results published by

Press (1965) and the plane-strain solution for the displacements due to dip-slip faulting derived by Freund and Barnett (1976). The latter deserve particular attention. Recall that we have already written down equations for the fault parallel displacement and shear strain due to an infinitely long vertical strike-slip fault which extends from the surface to depth, D. Similarly, Freund and Barnett (1976) found that the horizontal and vertical displacements, u_2 and u_3, and extensional strain, ε_{22}, due to slip on an infinitely long dip-slip fault are

$$u_2 = \frac{\Delta u \cos \theta}{\pi} \left[\tan^{-1}\left(\frac{x_2 - x_D}{D}\right) - \frac{(x_2 - x_P)D}{(x_2 - x_D)^2 + D^2} - \text{sign}(x_2)\frac{\pi}{2} \right]. \tag{22}$$

$$u_3 = \frac{\Delta u \sin \theta}{\pi} \left[\tan^{-1}\left(\frac{D}{x_2 - x_D}\right) - \frac{x_2 D}{(x_2 - x_D)^2 + D^2} \right.$$
$$\left. - \frac{\pi}{2}(1 - \text{sign}(x_2)) \right], \tag{23}$$

and

$$\varepsilon_{22} = \frac{\Delta u \cos \theta}{\pi} \left[\frac{2D(x_2 - x_D)(x_2 - x_P)}{\{(x_2 - x_D)^2 + D^2\}^2} \right], \tag{24}$$

where θ is the fault dip angle, $x_D = D/\tan \theta$, $x_P = D/\cos \theta \sin \theta = 2D/\sin 2\theta$, and u_3 is taken to be positive for uplift and negative for subsidence. Figure 5 shows the predicted vertical uplift and horizontal strain as a function of distance from the surface trace of a dip-slip fault.

The preceding equations give the surface deformations due to a long dip-slip fault. They are discussed in Singh and Rani (1992). Rani and Singh (1992) have generalized the results to an observation point at an arbitrary depth, and Singh and Rani (1993) discussed in some detail how the vertical and horizontal displacement patterns vary with fault dip angle and source depth.

There are many complexities in the real Earth that are not considered in the models discussed so far. Some obvious ones include the spherical rather than half-space shape of the Earth, vertical and lateral variations in the elastic constants, inelastic effects, variable slip, and geometric variations in the fault surface. We will discuss some of these issues below.

Note that the elastic constants of the Earth do not appear in some of the equations we have presented so far. In some cases this is a conse-

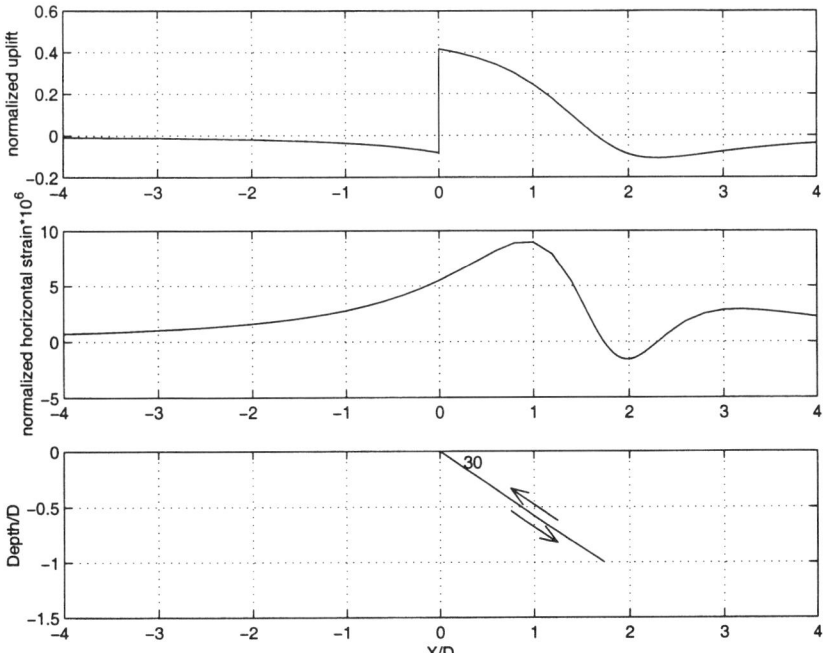

FIG. 5. Uplift and strain due to unit slip on a thrust fault extending from the surface to a depth of 30 km and with a dip of 30°.

quence of assuming an infinitely long fault, whereas in others there is an explicit assumption that Poisson's ratio = 1/4 (i.e., the two Lamé constants of elasticity theory are equal). Converse and Comminou (1975) showed that this assumption is generally a good one and is not a serious limitation even when Poisson's ratio is not 1/4.

Most of the attempts to fit geodetic observations of crustal deformation with analytical models have involved forward modeling, but work involving formal inversions has become more commonplace. One of the first researchers to formally invert geodetic data to derive earthquake source parameters was Jovanovich (1975), who used a least-squares procedure to invert strain observations for point source parameters describing the 1971 San Fernando, California earthquake and the 1969 Gifu, Japan earthquake. The inversion study of Matthews and Segall (1993) made use of the Volterra formula and allowed for depth-dependent slip. They showed that substantial coseismic slip may extend to greater depths than that estimated from uniform-slip models. For the 1906 San Francisco, California earthquake, they deduced that the coseismic slip may have extended to

15–20 km depth, whereas uniform-slip models give a faulting depth of only 10 km.

Abe (1977) published an interesting comparison of fault parameters derived from dislocation modeling with those deduced from tsunami records for a series of earthquakes in Japan in 1938. He concluded that the height of the sea-level disturbance at the margin of the tsunami source area agreed well with the slip deduced from dislocation modeling, although there was less than satisfactory agreement in the detailed features of the deformation pattern. The depth distribution of coseismic slip has been estimated from a combination of tsunami and geodetic data for the Nankai Trough region of Japan (Satak 1993) and for southern Alaska (Johnson et al., 1996).

2.3. Extensions beyond the Uniform Elastic Half-Space Model

2.3.1. Effects of Layering and the Use of Quasi-Static Matrix Propagator Techniques

The Earth is not a uniform elastic half-space. One of the first-order heterogeneities in Earth structure is a layering of the elastic properties with depth. Many of the analyses of elastic dislocations in a layered medium make use of the Thomson–Haskell propagator matrix. Here we will discuss some of the essential ideas of propagators and refer the reader to the books by Aki and Richards (1980) and Llibourty (1987) for more details. A particularly clear summary can be found in Ward (1985) and is abstracted here. The starting point is the equations of motion in the form as given by (4). We apply a Fourier transform to put this equation into the frequency domain. The Fourier transform of the function $u(r,t)$ is

$$U(r,w) = \int_{-\infty}^{\infty} u(r,t)\exp(i\omega t)\,dt, \qquad (25)$$

and the inverse transform is

$$u(r,t) = \frac{1}{2\pi}\int_{-\infty}^{\infty} U(r,\omega)\exp(-i\omega t)\,d\omega. \qquad (26)$$

The transformed equation of motion is

$$-\rho(r)\omega^2 U(r,\omega) = \nabla \cdot T(r,\lambda(\omega),\mu(\omega),\omega) - M(\omega)\cdot\nabla\delta(r-r_0), \qquad (27)$$

where T, the stress tensor, is a function of the elastic constants and the frequency. The second term on the right-hand side of the equation is the moment tensor representation of an earthquake point source located at r_0.

As before, the dynamic term on the left-hand side of the preceding equation can be ignored because we will be dealing with quasi-static problems; thus,

$$\nabla \cdot T(r, \lambda(\omega), \mu(\omega), \omega) = M(\omega) \cdot \nabla \delta(r - r_0). \quad (28)$$

As indicated, the Lamé constants, λ and μ, can be functions of frequency. We now assume that the problem under study is two-dimensional with layering of the material elastic properties in the z direction. We again apply a Fourier transform, but now to a spatial rather than a temporal dimension:

$$g(k, z) = \int_{-\infty}^{\infty} h(x, z) \exp(-ikx) \, dx. \quad (29)$$

With this transformation, the equilibrium equation can be written in the form

$$\frac{\partial}{\partial z} v(k, z) = A(k, x) v(k, z) + F(k, x). \quad (30)$$

In (30), v is a vector of the displacements and stress, A incorporates the elastic moduli, and F the moment tensor. For plane strain, where the displacement components are in the x–z plane,

$$v(k, z) = [iU_x(k, z), U_z(k, z), iT_{xz}(k, z), T_{zz}(k, z)]^T, \quad (31)$$

$$F(k, z) = \left[0, 0, ikM_{xx} + M_{xz}\frac{\partial}{\partial z}, ikM_{xz} + M_{zz}\frac{\partial}{\partial z}\right]^T \delta(z - z_0), \quad (32)$$

and

$$A(k, z) = \begin{bmatrix} 0 & k & \dfrac{1}{\mu} & 0 \\ \dfrac{k}{\lambda + 2\mu} & 0 & 0 & \dfrac{1}{\lambda + 2\mu} \\ \dfrac{4k^2\mu(\lambda + \mu)}{\lambda + 2\mu} & 0 & 0 & \dfrac{k\lambda}{\lambda + 2\mu} \\ 0 & 0 & -k & 0 \end{bmatrix}. \quad (33)$$

Similarly, for anti-plane strain (where the displacements are in the y direction),

$$v(k,z) = \left[U_y(k,z), T_{yz}(k,z)\right]^T, \quad (34)$$

$$F(k,z) = \left[0, ikM_{xy} + M_{yz}\frac{\partial}{\partial z}\right]^T \delta(z - z_0), \quad (35)$$

and

$$A(k,z) = \begin{bmatrix} 0 & \dfrac{1}{\mu} \\ \mu k^2 & 0 \end{bmatrix}. \quad (36)$$

With the aforementioned definitions and expressions, one can show that the displacements and stress obey an equation whose form is

$$v(k,z) = P(k,z,z_0)v(k,z_0) + \int_{z_0}^{z} P(k,z,\hat{z})F(k,\hat{z})\,d\hat{z}, \quad (37)$$

where P is the matrix propagator. The propagator's most important property is indicated by (37); it carries the solution from one depth to another. For a homogeneous layer, the anti-plane strain (strike-slip faulting) propagator is

$$P(k,z,z) = \begin{bmatrix} \cosh k\Delta z & (\mu k)^{-1}\sinh k\Delta z \\ (\mu k)\sinh k\Delta z & \cosh k\Delta z \end{bmatrix} \quad (38)$$

where $\Delta t = t - t_0$. The corresponding propagator for plane strain can be derived from the expressions in Ward (1984). One further step is required to utilize the propagator technique, i.e., the particular solution which satisfies the boundary conditions must be determined. This step constitutes a critical element of most papers which utilize the technique.

There have been many papers that have presented formulas for calculating the deformation due to faulting in a layered Earth. What follows is a brief summary. One of the first studies to consider the effects of layering was published by Singh (1970), who mainly considered point sources and presented explicit expressions for the displacements due to either strike-slip or dip-slip motion on a vertical fault. Sato (1971) used matrix methods to evaluate the effects of layering. He replaced various integrands that arise in this theory with approximations based on more analytically tractable functions. Sato's results were extended by Sato and Matsu'ura (1973) and Matsu'ura and Sato (1975), who considered the situation when the faulting extends over several layers in the half-space; however, a more explicit analysis of the effects of layering was published by Chinnery and Jovanovich (1972). They made use of results found in Rybicki (1971) and showed that the most important effect of having a weak layer next to the surface, in the case of slip on a long vertical strike-slip fault, is to enhance the displacement field. This result takes on added significance when one considers the effects of viscoelastic flow at depth, for this anelasticity produces a time-dependent reduction in effective rigidity. Another theory for the deformation due to a point dislocation in a half-space with an arbitrary number of layers was outlined by Jovanovich et al. (1974a, b). The authors compared the displacements due to point dislocations with those due to finite-size dislocations in a homogeneous half-space. They concluded that the two agree within 5% at distances greater than four times the fault length. They also showed that the presence of a buried weak layer alters the $1/(\text{distance})^2$ decay in the far-field displacement that occurs in a homogeneous Earth.

Several other papers considered specific problems within the context of layered Earth models. Mahrer and Nur (1979a), for example, modified the analytical theory for deformation due to slip on an infinitely long strike-slip fault embedded in a uniform medium to the case where the shear modulus increases monotonically with depth according to $\mu(z) = \mu_\infty[1 - \theta \exp(-z/z_0)]$, where θ and z_0 are numerical parameters. The computed surface displacements were affected much more for slip occurring on a surface rupture rather than for slip occurring on a buried fault. In a more recent analysis, Ma and Kusznir (1992) studied the three-dimensional displacements for a variety of configurations including sliding on an elliptical fault surface. A major aspect of their work was the computation of subsurface displacements. Ma and Kusznir (1994) extended the analysis from a layered elastic half-space to a layered elastic-gravitational half-space. In contrast to the work of Mahrer and Nur (1979a), they considered explicit examples where the elastic modulus decreases with depth. For a thrust fault, the depth weakening of the shear modulus results in an amplification of the maximum uplift in the overthrust block and maximum subsidence in the underthrust block. Roth (1990) extended the work of Singh (1970) to subsurface displacements in a layered elastic half-space. A useful appendix in his paper summarizes misprints from Jovanovich et al. (1974a) and Rundle (1978). Singh and Rani (1991) obtained closed-analytical expression for the displacements and stresses due to a two-dimensional source located in an isotropic half-space which lies over and is in welded contact with an anisotropic half-space. Savage (1998) expressed displacements in terms of potential functions and solved the resulting equations for an edge dislocation in a stratified elastic Earth consisting of a surface layer over a half-space. The most interesting physical result occurs when the surface layer is weaker than the half-space: the displacement field is similar to that of a uniform half-space but the dislocation appears to be deeper.

One of the most important results for a layered Earth comes from the case of an infinitely long strike-slip fault embedded in an elastic layer with thickness, H, and rigidity, μ_1, lying over a half-space with rigidity, μ_2. If the slip is uniform from the surface to a depth, D, then the displacement, V, is that due to a screw dislocation. For $x > 0$ (Rybicki, 1971),

$$V = \frac{V_0}{\pi} \left\{ \frac{\pi}{2} - \tan^{-1}\left(\frac{x}{D}\right) + \sum_{n=1}^{\infty} \left[\frac{\mu_1 - \mu_2}{\mu_1 + \mu_2}\right]^n \right. \\ \left. \cdot \left[\tan^{-1}\left(\frac{x}{2nH + D}\right) - \tan^{-1}\left(\frac{x}{2nH - D}\right)\right]\right\}. \quad (39)$$

As pointed out by Savage (1987), an implication of (39) is that the displacement field due to fault slip in a layer-over-half-space Earth is the same as that due to slip of a sequence of sources in a uniform half-space. The uppermost equivalent source is the same as that in the layered case, i.e., a source extending to depth D. At depths greater than H, however, there is a sequence of sources extending from depth $2nH - D$ to $2nH + D$. The normalized displacement amplitude for each of these sources is $[(\mu_1 - \mu_2)/\mu_1 + \mu_2)]^n$. This mathematical prescription provides an explanation for several of the physical effects discussed above, notably Chinnery and Jovanonich's (1972) finding that a weak subsurface layer enhances the displacements. Notice that when the rigidity of the lower layer (half-space) is less than that of the upper layer, i.e., $\mu_2 < \mu_1$, each term contributes a positive displacement, whereas when the upper layer is more rigid, the odd terms contribute a negative displacement. There is an important consequence for fitting theoretical results to observations. If one uses an elastic half-space model, the data may imply that there is negative slip at some depths when, in fact, the slip is positive but lies within a high-rigidity upper layer of a multilayer Earth.

2.3.2. Effects of Lateral Variations in Elastic Properties

Less attention has been devoted to the effects of horizontal variations in structure than to vertical stratification. However, some of the earliest studies, such as Sato (1974) and Sato and Yamashita (1975), considered the situation in which a fault is the boundary between rocks with different elastic constants. As expected, the surface displacements are amplified on the weaker rock. When the rigidity of the softer rock is one-half that of the harder rock, the computed maximum horizontal displacement is 1.6 times that for a uniform rheology. Rybicki and Kasahara (1977) and McHugh and Johnston (1977) also allowed the rigidity to vary laterally by making the fault zone weaker than the surrounding rock. The weakness of the fault zone causes the deformation to be concentrated in the near-field. Consequently, if a model with a uniform rheology model is used to estimate fault parameters in a situation in which the actual fault-zone rigidity is less than that of the surrounding rock, the fault depth will be underestimated. These lateral heterogeneity studies were extended somewhat by Rybicki (1978), who divided the half-space into several laterally distinct columns with different rheological properties. Mahrer and Nur (1979b) approached the problem of lateral variations in a similar manner to their aforementioned study of depth variations, i.e., they parameterized the lateral variation in rigidity with distance from the fault as $\mu(y) = [1 - \alpha \exp(-by)]\mu_\infty$, with α and b being model parameters. As in earlier studies, they found that the deformation was amplified in regions of low crustal rigidity. In subduction zones, one of the most important sources of heterogeneity is the presence of the subducting oceanic plate. We will discuss how it affects deformation after considering time-dependent processes.

2.3.3. Spherical Models

Although most numerical earthquake models have been based on a rectangular half-space, several workers have investigated the effects of sphericity. Ben-Menahem *et al.* (1970) and Ben-Menahem and Singh (1970) calculated the far-field displacements for both dip-slip and strike-slip sources. For strike-slip sources, they found that the absolute value of the ratio of the displacement at the surface of the sphere to those of a half-space varied from one to four. However, for shallow earthquakes, the differences in predicted coseismic displacements between the half-space and spherical models is only a few percent out to distances of 20° from the source. The effects of vertical or lateral heterogeneity may be far greater. Although most of the early studies did not include the effects of gravity, Sun and Okubo (1993) concluded that for a spherically symmetric, self-gravitating Earth model, the differences between the spherical and half-space results are less than 10% within 10° from the source. A similar conclusion was reached by Pollitz (1992, 1996), who used a spherical harmonic expansion for the displacements and concluded that within 100 km of an earthquake, the effects of ignoring sphericity are less than 2%. At distances of 1000 km or more from the source, however, the effects of sphericity may be important, and although the predicted displacements are small, they are potentially detectable by modern geodetic techniques provided competing local effects can be accounted for.

Okubo (1993) considered the displacement due to a point source in a spherical Earth and contrasted the displacements due to an earthquake with those due to tidal forces, surface loading, and surface shear. He showed that the coseismic displacements can be written in terms of a tidal deformation, a radial displacement due to pressure, and a tangential displacement due to shear and torsion. This work was updated by Sun et al. (1996), who derived the displacement field for a point source embedded in a spherically layered elastic earth. They calculated vertical and horizontal coseismic displacements for the 1964 Alaska earthquake using an elastic parameter model developed by Gilbert and Dziewonski (1975). Horizontal displacements were 1 cm at 30° from the source and 0.5 cm at 40° from the source. The globally averaged radial displacement was 0.025 cm. Advances in geodetic techniques since the 1960s have made displacements of a few millimeters potentially detectable.

2.4. The Finite-Element Method

Once the fault slip and geometry are specified, the Earth's rheology is given, and certain boundary conditions are met, it should be possible to calculate the displacements due to an earthquake at all locations. Numerical techniques provide methods of carrying out this calculation even when the appropriate Green's functions are not known or the required integrations cannot be performed. The two most widely used numerical techniques are finite-element and boundary-element techniques. Because of the far greater utility and frequency of use in problems of varying rheology, we will focus on finite-element techniques. For boundary-element techniques, the reader is referred to such texts as Beer and Watson (1994). The finite-element technique is based on the following approach:

(1) Discretize the body under study by representing it through a mesh of elements and nodes as shown in Figure 6. The body may be irregularly shaped, but the shape is approximated by a set of elements which have simple geometries such as the rectangles and triangles that are shown in the figure. The material properties may vary from element to element.

(2) Within each element, represent the displacement as a simple interpolation function based on the displacements at the nodes. For example, in one dimension, the linear approximation for the displacement, u, between nodes at $x = 0$ and $x = L$, is $u = [1 - x/L]u_1 + [x/L]u_2$, where u_1 and u_2 are the displacements at the nodes $x = 0$ and L, respectively. The extensional strain within the element, $e = [u_2 - u_1]/L$, is a constant.

(3) Utilize the force equilibrium condition, say in the form of the Principle of Virtual Work, and the constitutive law for the material to derive a set of algebraic equations for the nodal displacements.

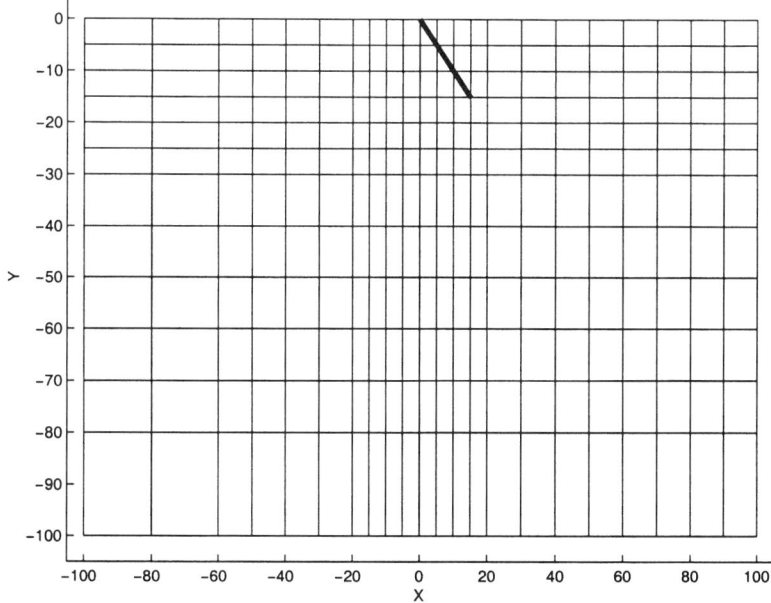

FIG. 6. A two-dimensional finite-element grid based on rectangular and triangular elements. The fault is represented by the heavy line.

(4) Apply the appropriate boundary conditions in the form of applied forces or known boundary nodal displacements.

(5) Invert the algebraic equations to solve for the unknown nodal displacements.

We now repeat these steps in some detail and in a slightly different order. Again consider an elastic medium, although not necessarily a homogeneous one. The matrix relationship between stress and strain is $\sigma = D\varepsilon$, where D is the material matrix. If the object under study is divided into a finite number of subdivisions or elements, the stress–strain relationship is valid for each element, m, as well as for the whole. Thus we write

$$\sigma^m = D^m \varepsilon^m. \tag{40}$$

Assume that the system under study is an equilibrium and then subject it to an infinitesimal virtual displacement. The Principle of Virtual Work requires that the internal virtual work equal the external virtual work, or

$$\int_V \Delta \varepsilon \sigma \, dV = \int_V \Delta u f^B \, dV, \tag{41}$$

where $\Delta\varepsilon$ is the virtual strain, f^B is the applied body force, and V is the body volume. The preceding equation can be generalized to take into account forces other than body forces. For example, an applied surface force will also appear on the right-hand side of the preceding equation in a surface integral. After discretizing the volume into elements, the preceding equation becomes

$$\sum_m \int_{V^m} \Delta\varepsilon^m D^m \varepsilon^m \, dV^m = \sum_m \int_{V^m} \Delta u f^B \, dV^m. \tag{42}$$

Now assume that the displacement in the interior of each element can be written in terms of the nodal displacements, so

$$U^m(x, y, z) = N^m(x, y, z) U, \tag{43}$$

where $N(x, y, z)$ is an interpolating function with an analytically simple form; in one dimension it is often a linear or quadratic polynomial, while in two dimensions it is usually a bilinear or biquadratic form. Because the interpolation function is not exact, (43) should be written with an approximately equal sign, but the equal sign is used for simplicity. Differentiating the preceding equation provides a relationship between the strain and nodal displacements:

$$\varepsilon^m(x, y, z) = B^m(x, y, z) U, \tag{44}$$

where B is the generalized symbol for spatial derivatives of N. A key feature of the finite-element approach is that the differentiation is applied to the interpolating polynomial whose functional form is known and whose derivatives are easily calculated, rather than to the exact, but unknown, functional form for the displacements. The next step is to substitute (44) into (42) to derive

$$\Delta U^T \left[\sum_m \int_{V^m} (B^m)^T D^m B^m \, dV^m \right] U = \Delta U^T \left[\sum_m \int_{V^m} N^m f^B \, dV^m \right]. \tag{45}$$

The term under the integration sign in the left-side bracket is determined by the rheological properties of the material, the form chosen for the displacement interpolation function, N, and the discretization. This function is called the local or element stiffness matrix, k^m. The integral in the right-side bracket is the applied force acting on the body element. The virtual displacement vector is arbitrary; therefore, it follows that

$$\left[\sum_m k^m \right] U = \sum_m F^m, \tag{46}$$

where

$$k^m = \int_{V^m} (B^m)^T D^m B^m \, dV^m \qquad (47)$$

and

$$F^m = \sum_m \int_V N^m f^B \, dV^m. \qquad (48)$$

The global stiffness matrix is $K = \sum_m k^m$ and $F = \sum_m F^m$, so we end up with the simple matrix equation

$$KU = F, \qquad (49)$$

which can be inverted to give the nodal displacements. Equation (49) is the basic equation of finite-element analysis.

One important subtlety associated with the application of the finite-element method just outlined concerns the application of the fault-slip constraint. This discontinuity manifests itself as a constraint on the relative slip rather than absolute displacement on the fault. There are several techniques for dealing with this important boundary condition. One is to utilize continuity of stress across the fault. The displacement discontinuity method utilizes a multinode element of zero thickness to represent the fault. The element is given various stiffness parameters to allow for a variety of conditions. For example, slip can be imposed by giving the element zero stiffness in shear. An efficient alternative is the split-node technique developed by Melosh and Raefsky (1981). Suppose the fault passes through node i and separates elements r and s. Rather than assuming continuity in displacement, we write the displacement for node i as $u^r = u - \Delta u$ when the nodal displacement is evaluated in element r and $u^s = u + \Delta u$ when the nodal displacement is evaluated in element s. The displacement discontinuity or fault slip is $2\Delta u = u^r - u^s$. Here, u is, of course, the mean displacement. Terms in the finite-element equation involving the product of the stiffness times the slip, Δu, can be brought over to the known right-hand side of the finite-element equation as an effective force, leaving the mean displacement as one of the unknowns to be determined. While the split-node technique is sufficient for modeling fault slip in many situations, it can also be generalized to allow for fault slip with a prescribed degree of resistance. This is the slippery-node method and is discussed in Melosh and Williams (1989).

Up to this point we have assumed an elastic rheology. When viscoelastic effects are considered, there is a further complication that arises due to the time dependence in the stress–strain relationship. This issue will be addressed after we discuss viscoelasticity. Texts on finite-element analyses (e.g., Zienkiewicz, 1977; Bathe, 1982) discuss techniques for choosing the interpolation functions, methods for evaluating the surface and volumetric integrals that appear in the equations, and techniques for solving the coupled linear algebra equations. Seemingly prosaic issues such as numbering the nodes and elements can actually be quite important because they can affect the efficiency of the numerical algorithms (Akin, 1982).

3. Time-Dependent Effects

A common observation in the aftermath of significant earthquakes is that the deformation is rapid compared to either the preseismic rate or the interseismic average. One mechanism for such accelerated deformation is further seismic or aseismic sliding on the coseismic rupture plane, its lateral extension, or a subsidiary fault. Another proposed mechanism for rapid postseismic deformation is the flow of crustal fluids in response to the change in fluid pressure caused by the earthquake (Nur and Booker, 1972). Two deeper-seated, time-dependent processes are particularly important mechanisms of postseismic rebound and are the focus of this section. These are fault creep occurring down-dip of the coseismic rupture plane and viscoelastic shear flow in the ductile portions of the lower crust and upper mantle. Both are stimulated by the coseismic transfer of stress from the shallow seismogenic portions of the Earth to greater depths.

3.1. Deep Fault Creep

With regard to deep fault creep, we recognize that because temperatures are higher in the lower crust and upper mantle than in the shallow seismogenic layers, the deeper portions of a fault may respond to stress changes by stable sliding rather than by brittle failure. The surface deformations produced by the sliding can be calculated from dislocation theory, although the time dependence of the fault slip is usually determined empirically. Thus, the equations that we have discussed above in connection with coseismic displacements can be employed to model postseismic deformation simply by taking the fault plane to lie at depths appropriate for the postseismic sliding and allowing the slip to be time-dependent. As in the coseismic case, complex slip distributions or geometries can be approximated by breaking the distributions into small segments each with uniform properties, and then adding the contributions of each segment to the total deformation pattern. One example, among many others, of modeling postseismic uplift in terms of creep at depth can be found in

Brown *et al.* (1977), who examined several epochs of leveling data collected during the decade following the aforementioned 1964 Prince William Sound, Alaska earthquake. The data, shown in Figure 7a, indicate postseismic uplift centered a few hundred kilometers landward of the Aleutian trench, the surface boundary between the North American and Pacific plates. The rate of uplift decreases with time elapsed since the earthquake. Brown and co-workers attributed the uplift through 1975 to 2.3 m of fault creep on a down-dip, steepened extension of the coseismic rupture plane (Fig. 7b). Although some of the geometric parameters of this model, notably the dip angle of the creeping plate interface, should be changed in light of current information, the general agreement between the observations and theory was an early success for the deep-slip model.

3.2. Viscoelastic Flow

Another deep-seated mechanism for postseismic deformation is viscoelastic flow. Experimental work suggests that rocks in the lower crust and the mantle should undergo solid-state flow in response to applied shear stress. The coseismic transfer of stress from the seismogenic layer to deeper layers of the Earth is accommodated initially by elastic deformation and subsequently by shear, ductile solid-state flow. Classically, the viscoelastic response is confined to the asthenosphere, but as Turcotte *et al.* (1984) point out, rock mechanics experiments also suggest that viscoelastic deformation can occur at intermediate and lower crustal depths. We can represent the viscoelastic behavior by a Maxwell body consisting of two of the linear elements shown in Figure 8, namely, an elastic spring and a viscous dashpot connected in series. The stress is the same in the two elements while the strain for the entire body is the sum of the strains for the individual elements. For the elastic element, the relationship between shear stress, σ, and strain, ε, is, as before, $\sigma = 2\mu\varepsilon$, while for the viscous element, $\sigma = 2\eta\dot{\varepsilon}$, where η is the viscosity. The constitutive equation for the Maxwell body is, therefore,

$$\dot{\varepsilon} = \frac{\dot{\sigma}}{2\mu} + \frac{\sigma}{2\eta}. \tag{50}$$

Equation (50) describes the shear behavior of the Earth, whereas the volumetric behavior is nearly elastic. An important parameter associated with viscoelastic flow is the Maxwell time, $\tau = \eta/\mu$. Unfortunately, various authors use slightly different forms for (50) and the Maxwell time. In general, the Maxwell time is defined as $\tau = \alpha(\eta/\mu)$, where α is 1/2, 1, or 2. Notice that when the strain is constant, then the stress decays to $1/e$ of

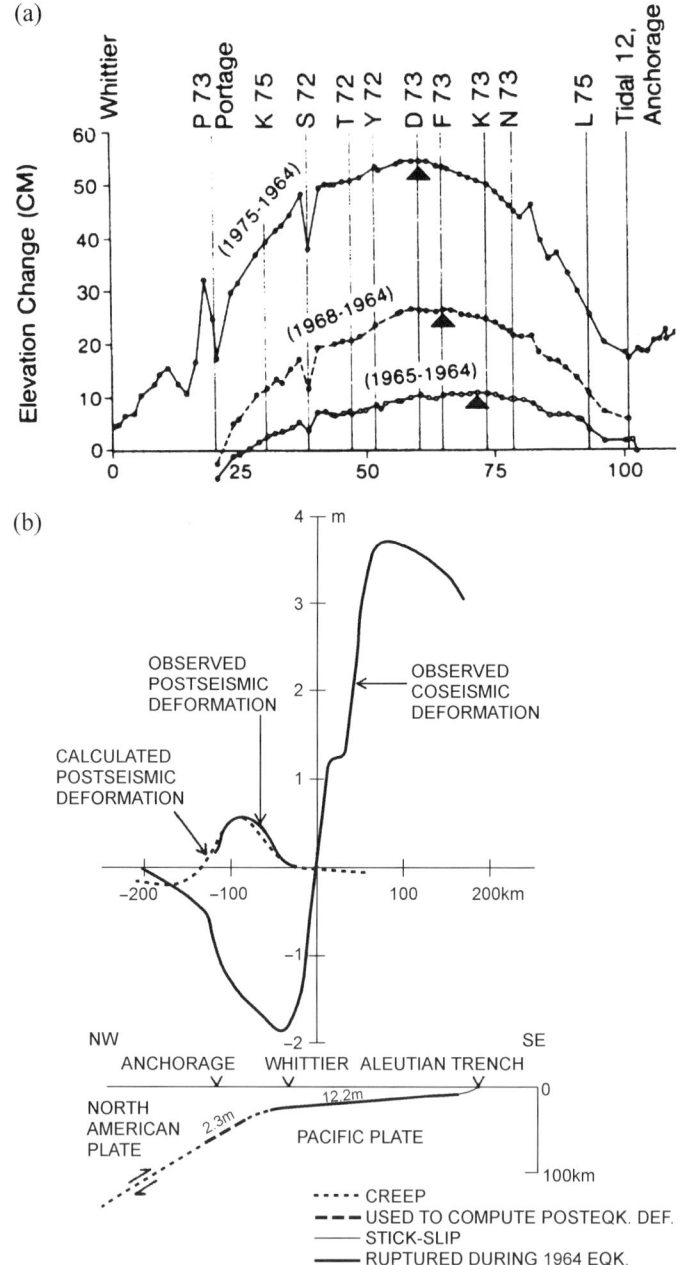

FIG. 7. (a) Uplift determined by repeated leveling surveys along Turnagain Arm, Alaska following the 1964 Prince William Sound earthquake. (b) Deep-creep model for the cumulative (1975–1964) uplift. (From Brown et al. 1977.)

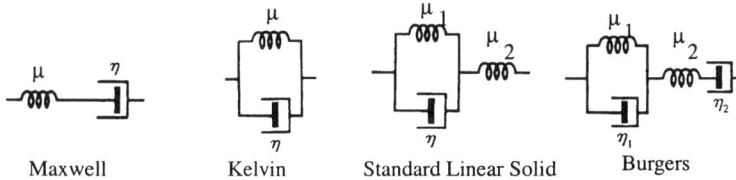

FIG. 8. Four (linear) viscoelastic models. The Maxwell body consists of an elastic and a viscous element connected in series; the Kelvin body is an elastic and a viscous element in parallel. One representation of the standard linear solid is a series connection of elastic and Kelvin elements. The Burgers body is a series combination of Maxwell and Kelvin bodies. The strains are additive for elements in series, while the stresses add for elements in parallel.

its initial value at $t = \tau$ ($\alpha = 1$); however, in the Earth both the strain and stress change with time, so the postseismic viscoelastic flow has a more complex temporal behavior as illustrated below. Equation (50) is one example of a linear constitutive law. Others may be obtained by adding additional elements and by arranging the elastic and viscous elements in parallel, rather than in series. For parallel elements, the strains in the two elements are equal and the stresses add. The Kelvin body, for example, consists of a spring arranged in parallel with a dashpot, while the standard linear solid consists of a spring connected in parallel to a Maxwell element (or, equivalently, a spring in series with a Kelvin element). The standard linear solid is of particular interest because it responds elastically to a sudden stress, then relaxes a portion of the stress while retaining some permanent rigidity. A general form of (50) which incorporates several different rheologies is (Jaeger and Cook, 1976)

$$\left(a_0 + a_1 \frac{\partial}{\partial t}\right)\sigma_{ij} = \left(b_0 + b_1 \frac{\partial}{\partial t}\right)\varepsilon_{ij}. \qquad (51)$$

3.2.1. Correspondence Principle

We have seen that for the linear viscoelastic constitutive relation, i.e., (51), the left-hand side of the equation consists of a linear operator applied to the stress and the right-hand side consists of a linear operator acting on the strain. The Correspondence Principle is a very useful tool for deriving the deformation for subsidences with linear rheologies, particularly a layered Earth, when the solution for the corresponding elastic problem is known. It is based on the observation that the relationship between the Laplace transform of stress and the Laplace transform of strain takes the same form for all linear rheological laws. We form the Laplace transform of (51) by multiplying both sides of the equation by $\exp(-st)$, where s is the Laplace transform variable, and integrating over time from $t = 0$ to

$t = \infty$. Explicitly,

$$\int_0^\infty \left(a_0 + a_1 \frac{\partial}{\partial t}\right) \sigma_{ij}(t) e^{-st} \, dt = \int_0^\infty \left(b_0 + b_1 \frac{\partial}{\partial t}\right) \varepsilon_{ij}(t) e^{-st} \, dt. \quad (52)$$

The Laplace transform of time-dependent variable, $g_{ij}(t)$, is

$$L[g(t)] = \bar{g}(s) = \int_0^\infty g(t) e^{-st} \, dt. \quad (53)$$

Making use of the fact that $L[g'(t)] = s\bar{g}(s) - g(0^+)$, we find

$$\bar{\sigma}_{ij}(s) = \frac{b_0 + b_1 s}{a_0 + a_1 s} \bar{\varepsilon}_{ij}(s), \quad (54)$$

where the prime symbol indicates time differentiation. Defining

$$\bar{\mu}(s) = \frac{1}{2} \frac{b_0 + b_1 s}{a_0 + a_1 s}, \quad (55)$$

we find

$$\bar{\sigma}(s) = 2\bar{\mu}(s) \bar{\varepsilon}(s), \quad (56)$$

that is, the Laplace transformed viscoelastic equation has the same form as the Laplace transformed elastic equation provided that the shear modulus is replaced by the appropriate $\bar{\mu}(s)$. Since the viscoelastic material responds to the same quasi-equilibrium requirements as the purely elastic material, the solution to the viscoelastic problem can be generated once the elastic solution is known. To get the viscoelastic solution, one proceeds as follows:

(1) Replace the shear modulus in the elastic solution by its Laplace transform; similarly, replace the slip time function by its Laplace transform. For coseismic slip, the time function is just the slip, V_0, for $t \geq 0$. Its Laplace transform is V_0/s. If the fault is creeping at a constant velocity, the Laplace transformed slip function is only slightly more complicated, as we will see in Section 3.1.1.

(2) Compute the inverse transformation.

As an illustration, we review the model of postseismic rebound developed by Nur and Mavko (1974) as applied to a vertical strike-slip fault. The fault extends from the surface to depth D in an elastic layer of thickness, H. The elastic layer is underlain by an elastic half-space. The fault is infinitely long in the third dimension, so the problem is one of anti-plane or out-of-plane strain, a screw dislocation problem. We have already presented in (39) an expression for the surface displacements, V, namely,

$$V = \frac{V_0}{\pi} \left\{ \frac{\pi}{2} - \tan^{-1}\left(\frac{x}{D}\right) + \sum_{n=1}^{\infty} \left[\frac{\mu_1 - \mu_2}{\mu_1 + \mu_2}\right]^n \right. \\ \left. \cdot \left[\tan^{-1}\left(\frac{x}{2nH + D}\right) - \tan^{-1}\left(\frac{x}{2nH - D}\right)\right] \right\}. \quad (57)$$

We now replace the elastic half-space with a standard linear solid half-space. The upper layer is still elastic, but now it is underlain by a viscoelastic asthenosphere. Accordingly, we replace μ_2 by the corresponding Laplace transform modulus for a viscoelastic body as given by (55) and replace V_0 by V_0/s. Once these replacements have been inserted into (57), we compute the inverse Laplace transform to obtain the solution to the viscoelastic problem. Although Nur and Mavko (1974) used a standard linear solid for the viscoelastic asthenosphere, the results can be expressed in a particularly compact form when the asthenosphere is a Maxwell substance and the elastic rigidities of the lithosphere and asthenosphere are equal. In this case, the displacement equation can be written

$$\frac{V}{V_0} = \frac{1}{2} - \frac{1}{\pi} \tan^{-1}\left(\frac{x}{D}\right) + \sum_{n=1}^{\infty} V_{ve}(x, t, n), \quad (58)$$

where $V_{ve}(x, t, n) = G(x, n) \cdot T(t, n)$, and

$$G(x, n) = \frac{1}{\pi} \tan^{-1}\left\{ \frac{2\frac{x}{D}}{\left(2n\frac{H}{D}\right)^2 - 1^2 + \left(\frac{x}{D}\right)^2} \right\}, \quad (59)$$

$$T(t, n) = \frac{1}{(n-1)!} \int_0^{t/2\tau} r^{n-1} \exp(-r)\, dr. \quad (60)$$

T is the incomplete Gamma function. The $n = 1$ term is simply

$$T(t, 1) = 1 - \exp\left(-\frac{t}{2\tau}\right). \quad (61)$$

Subsequent terms can be derived from the recursion relation:

$$T(t, n+1) = T(t, n) - \frac{r^r \exp(-r)}{n!}, \quad (62)$$

where $r = t/2\tau$ and $\tau = \mu/\eta$ is again the Maxwell time of the astheno-sphere (see also Savage and Prescott, 1978a, but note that their definition of the asthenosphere relaxation time is twice that used here). The T's all approach 1 in the limit of infinite time. Figure 9 shows the spatial variation of G and the temporal variation of T for $n = 1-4$. Some interesting observations include:

(a) The G's peak at $x = \sqrt{(2nH)^2 - D^2}$. For $H = D$, the peak amplitude of the $n = 1$ term is exactly $1/6$ and occurs at $x = \sqrt{3}D$.

(b) In the near-field, only the first few terms in the series expansion contribute substantially to the deformation, while in the far-field many more terms are required. Furthermore, for the higher-order terms, it takes longer for the temporal function to approach 1. This is essentially a temporal-spatial decomposition of diffusion.

(c) Only the $n = 1$ terms has a purely exponential temporal relaxation, although the higher-order terms also contain exponential factors.

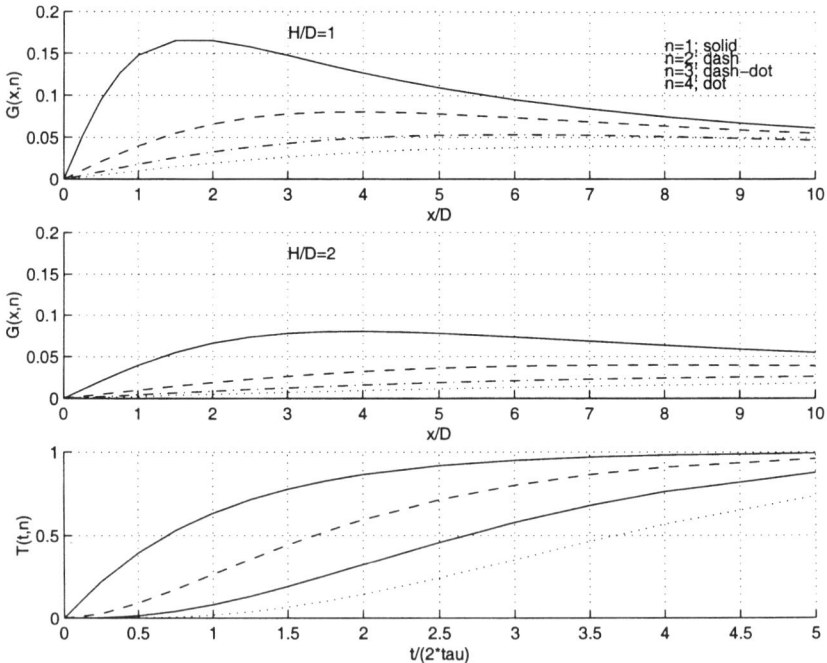

FIG. 9. The spatial, $G(x, n)$, and temporal, $T(t, n)$, functions occurring in a modified version of the Nur and Mavko (1974) model of viscoelastic postseismic rebound in which the asthenosphere is a Maxwell body with rigidity equal to that of the overlying elastic lithosphere. The top two panels show the spatial function for different values of the ratio of lithospheric thickness to fault depth. The bottom panel shows the temporal function.

Nur and Mavko (1974) found that the effect of viscoelastic flow in the asthenosphere is to enhance the fault parallel displacements postseismically. An example, for the aforementioned case of a Maxwell asthenosphere, is shown in Figure 10. The figure shows the cumulative postseismic displacement as a function of distance from the fault for several values of the ratio of fault depth to elastic layer thickness. The displacements are in the same direction as the coseismic displacements, and as one might expect, become more significant as successively greater fractions of the lithosphere are ruptured. For $D/H = 1$, the cumulative coseismic plus postseismic displacement ($t = \infty$) approaches the fault slip at all distances from the fault. Nur and Mavko (1974) also attempted to find a solution to the corresponding solution for an edge dislocation, i.e., the dip-slip problem; however, the later analysis involves further approximations in the elastic solution which appear to lead to erroneous results (see, for example, Thatcher and Rundle (1979) and Singh and Rani (1994) for discussions of this point).

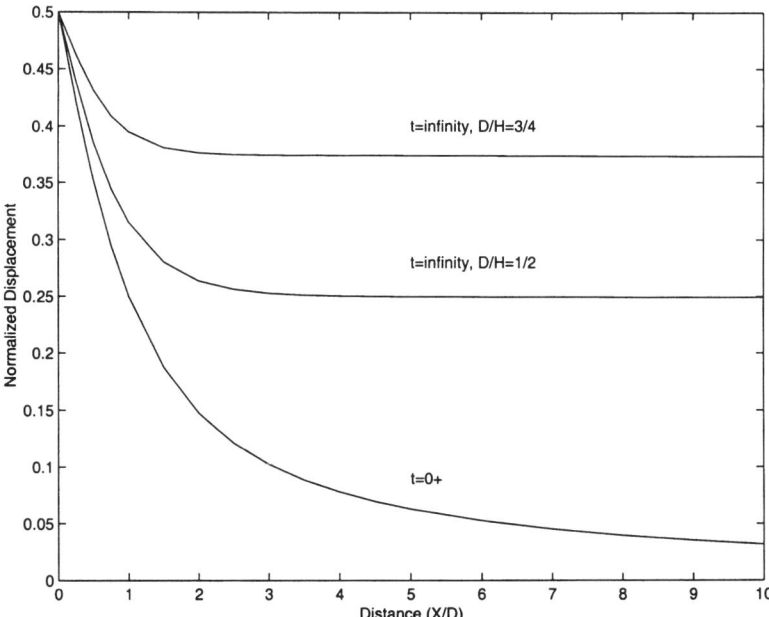

FIG. 10. Postseismic displacement versus distance for viscoelastic rebound following slip on a strike-slip fault.

3.2.2. Viscoelasticity and the Finite-Element Method

Viscoelasticity introduces new terms into the finite-element equations. Following Zienkiewicz (1977), an approach to solving the quasi-static equilibrium equations is as follows. First, assume that stress is instantaneously elastic, i.e.,

$$\sigma = DE = D(\varepsilon - e), \tag{63}$$

where E and e are, respectively, the elastic and viscous contributions to the total strain, ε. It follows that the change in stress between times m and $m + 1$ is

$$\sigma_{m+1} - \sigma_m = D[\varepsilon_{m+1} - \varepsilon_m] - D[e_{m+1} - e_m]. \tag{64}$$

Assuming that the stress is not changing too rapidly, we use the constitutive equation for the viscous element, $\dot{e} = \sigma/2\eta$, to write the approximation

$$e_{m+1} - e_m = (\sigma \Delta t)/(2\eta) = \beta(\sigma_m)\Delta t. \tag{65}$$

Note that in this approximation, $\beta = \sigma/2\eta$ is evaluated at time m. Using the usual finite-element strain-displacement matrix, B, (64) can be rewritten as

$$DB[U_{m+1} - U_m] = \sigma_{m+1} - \sigma_m + D\Delta t \beta_m. \tag{66}$$

This expression forms the basis for an iterative scheme. Multiply both sides by B^T and integrate over volume. Then for the zeroth iteration,

$$\left[\int B^T DB \, dV\right] \Delta U^0_{m+1} = \int B \Delta \sigma^0_{m+1} \, dV + \int B^T D\Delta t \beta_m \, dV. \tag{67}$$

The condition of quasi-static equilibrium (equivalent to the Principle of Virtual Work) requires

$$\int B^T \sigma \, dV + F = 0, \tag{68}$$

from which

$$\left[\int B^T DB \, dV\right] \Delta U^0_{m+1} = -\Delta F + \int B^T D\Delta t \beta_m \, dV. \tag{69}$$

This equation resembles the earlier finite-element equation, the major difference being that there is now an additional term due to the viscous flow. The prescription for implementing the viscoelastic aspects of the finite-element method begins with solving the initial elastic problem to get the displacements. The displacements are used to calculate the stresses, then (69) is used to calculate the zeroth-order change in the displacements at the first time increment. These displacements are used to calculate the new stresses and, in turn, the new stresses are used to calculate a correction to the displacements. The process is continued in an iterative manner until convergence is achieved. An assumption in the procedure just outlined is that

$$e_{m+1} = e_m + \beta(\sigma_m)\Delta t, \tag{70}$$

that is, the viscous correction is calculated at time m. This assumption restricts one to taking small time steps, particularly if the viscous term is changing rapidly with time. However, a more general expression is

$$e_{m+1} = e_m + \beta(\sigma_{m+\delta})\Delta t, \tag{71}$$

where

$$\sigma_{m+\delta} = (1-\delta)\sigma_m + \delta\sigma_{m+1}. \tag{72}$$

The case $\delta = 0$ corresponds to the earlier fully explicit integration and the case $\delta = 1$ corresponds to a fully implicit integration. Implicit integration is more computationally intensive than explicit integration, but is numerically stable for larger time steps.

3.2.3. Nonlinear Viscoelasticity

The linear relationship between stress and strain rate that we have assumed for a viscous body in the previous sections does not necessarily conform to the behavior of real rocks as inferred from laboratory experiments. While there are mechanisms such as diffusion creep, i.e., the diffusion of an impurity in the rock crystal lattice or the diffusion of a lattice vacancy point defect, which do obey linear flow laws, others do not. Many of the mechanisms for viscous deformation are thermally activated and obey a solid-state creep law of the form

$$\dot{\varepsilon} = A \exp\left(-\frac{E+PV^*}{RT}\right)\sigma^n, \tag{73}$$

where E and V^* are the activation energy and the activation volume, respectively, R is the gas constant, and T is temperature. Equation (73) is the uniaxial version of a more general multidimensional constitutive equation. For $n = 1$, we get the linear flow law; however, for typical lower crust or upper mantle rocks, $n \approx 3$ with both higher and lower values having been observed. Dislocation creep, a process wherein dislocations move through the rock lattice, is a common cause of ductile deformation for which $n \approx 3$. Despite the fact that most rocks exhibit a nonlinear relationship between strain rate and stress for the conditions found in the lower crust and upper mantle, most modeling efforts have used a linear relation. There are several reasons for this:

(1) For purely analytical models, there are very few solutions which can be obtained using a nonlinear rheology. By contrast, the Correspondence Principle provides a straightforward way to generate the viscoelastic solution provided the corresponding solution to a layered elastic medium is already known.

(2) For numerical models, linear rheologies are less computationally demanding because the temporal integration does not require the iterative procedures demanded by nonlinear rheologies.

(3) Linear viscosity is useful for contrasting the dissipative behavior of the deforming medium at different frequencies including those associated with seismic wave propagation (seconds), postseismic rebound (years to decades), postglacial rebound (10^3–10^4 yr), and mantle convection (10^6 yr).

Although linear rheologies have been employed in the vast majority of numerical simulations, the use of nonlinear rheologies is becoming more common as computational power increases.

3.2.4. Viscoelastic Postseismic Rebound

Singh and Rosenman (1974) and Rosenman and Singh (1973a, b) studied deformation and stress relaxation due to faulting on a rectangular vertical plane in a uniform Maxwell viscoelastic half-space. Although current interest in such models is limited because the uniform viscoelastic half-space is not a very good representation of the Earth on the time scale of the earthquake cycle, the results are somewhat instructive from an academic perspective. They found that the surface displacements due to slip on a vertical dip-slip fault are the same for viscoelastic and elastic half-space rheologies. For vertical strike-slip faults, however, the displacements do depend on rheology.

Matsu'ura and Tanimoto (1980) considered the response of a uniform viscoelastic half-space to slip for a more general fault geometry and found that the transient postseismic displacement, when present, has a time dependence: $1 - \exp(-t/\tau)$. As might be expected from the aforementioned results of Singh and Rosenman (1974), only the strike-slip components contribute to the postseismic motion for a vertical fault. For a horizontal fault, all postseismic terms vanish.

Following the work of Nur and Mavko (1974), many researchers developed models of postseismic rebound in which the rebound was attributed to viscoelastic flow in a lower crust or upper mantle that lies under an elastic lithosphere. Rundle and co-workers have been instrumental in developing a propagator approach to modeling the earthquake cycle. The mathematical development of their model can be found in Rundle (1976, 1978, 1980, 1982), and Rundle and Jackson (1977a). The starting point for this theory is the use of the Haskell–Thomson matrix propagators as expressed by Singh (1970) and Jovanovich et al. (1974a). Viscoelasticity is introduced through the Correspondence Principle. Much of the work involved in developing the model involves computation of the inverse Laplace transforms; however, in later papers by Rundle and co-workers, the inversion is simplified by using the Prony series in which the temporal behavior of the displacement is expanded in terms of a series of decaying exponentials, i.e., $u(t) \approx \Sigma_{ij}^{N} A_i \tau_j (1 - \exp(-t/\tau_j))$, where the A_i are a set of constants that are determined by least-squares techniques. The τ_j are relaxation times, for example, the set $\{\tau, 2\tau, 10\tau, 20\tau, 100\tau, 200\tau\}$, where $\tau = \eta/\mu$ is again the Maxwell time for the asthenosphere. One of the important aspects of the work by Rundle and his colleagues has been the consideration of gravitational effects. They show, for example, that gravitational effects are usually minimal in the short term after an earthquake, but become significant later on (e.g., Fernandez et al. 1996). Figure 11 shows the vertical displacements due to postseismic rebound from thrust faulting where only the upper half of the elastic lithosphere is ruptured during the earthquake. The dominant characteristic is subsidence over the thrust, but notice how the subsidence is reduced when gravitational forces are considered. Rundle (1982) found that the vertical displacements computed from purely viscoelastic solutions are still changing significantly after 10 Maxwell times, whereas those computed from viscoelastic-gravitational solutions have almost reached equilibrium.

Rundle and Jackson (1977b) applied their formulation to the San Francisco earthquake of 1906. They found that a variety of models which differ in the amount of aseismic slip, viscoelastic relaxation, and steady-state plate motion could fit the available geodetic data. Ambiguity, both in the magnitude of model parameters and in the determination of the dominant mechanism responsible for crustal deformation, is a common problem in studying crustal deformation, particularly when the observational constraints are provided only by surface geodetic data. In fact, Savage (1990) has pointed out that for a long, vertical, strike-slip fault, there is a mathematical equivalence between viscoelastic rebound and elastic slip-at-depth models. The surface deformation pattern due to viscoelastic flow can be replicated by imposing creep on a distribution of slip sources lying on a vertical plate in an elastic half-space. The ambiguity points out the need to combine geodetic observations with constraints

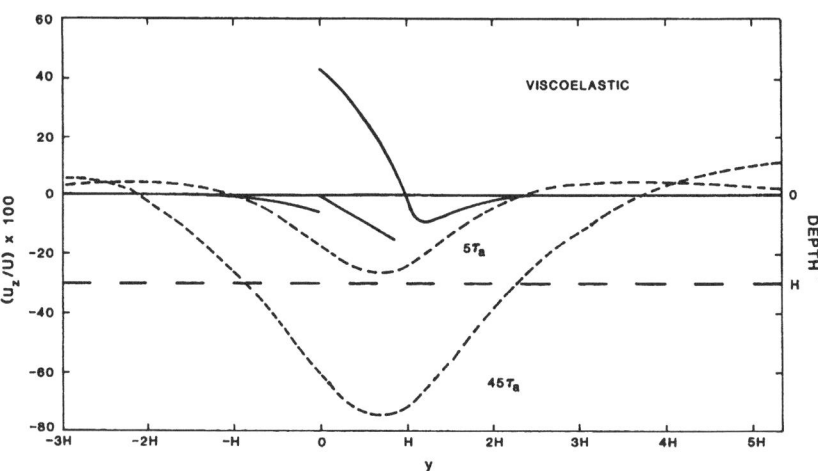

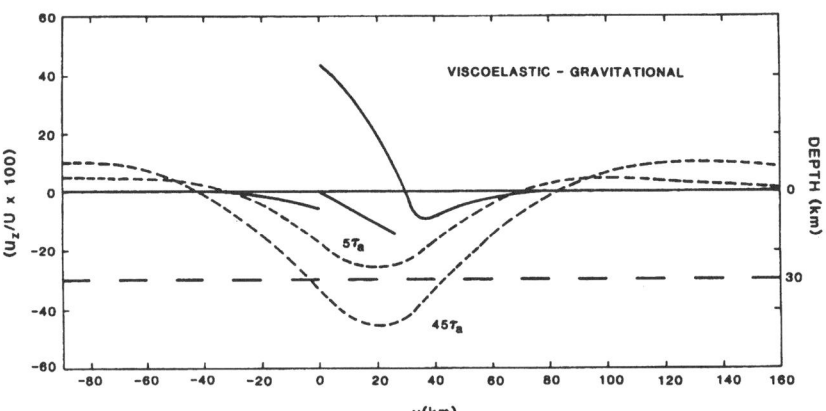

Fig. 11. Vertical postseismic motion due to viscoelastic rebound following a thrust earthquake. The fault dip angle is 30° and the fault length is $2H/3$, where H is the elastic lithosphere thickness. The coseismic motion is shown by the solid line, while postseismic motion is shown at $t = 5\tau_a$ and $t = 45\tau_a$, where $\tau_a = 2\eta/\mu$. (a) Viscoelastic rebound only, no gravitational forces. (b) Viscoelastic rebound plus gravitational forces. (From Rundle 1982.)

from seismology and geology in determining the appropriate physical mechanism and model parameters in specific environments.

The reader of Rundle's papers will discover that the Lamé constant of the asthenosphere, λ, is held elastic while, as usual, the shear modulus, μ, is allowed to relax viscoelastically. The unusual aspect of dealing with the rheology in this manner is that the bulk modulus undergoes some viscoelastic relaxation. This is probably not a major concern for dealing with those fault problems where the driving mechanism involves shear, rather than volumetric, stress. Numerical experiments by Rundle (1982) indicate that displacements are changed by only a few percent for typical fault problems; however, there are situations where volumetric dilatations and contractions are significant and the disagreements may be larger.

Matsu'ura *et al.* (1981) and Iwasaki and Matsu'ura (1981) obtained equations for the displacements, strains, and tilts due to rebound following faulting in a multilayer half-space where one of the layers is viscoelastic. They allowed for arbitrary dip angle, depth of fault, and direction of slip and included the possibility that the faulting was in the viscoelastic layer. They obtained explicit solutions for point sources and used numerical integration to evaluate displacements over a rectangular fault. For some illustrative cases, they found that the near-field postseismic displacements are nearly complete after 10 Maxwell times, a result in accord with Rundle (1982).

Melosh (1976) developed an analytical model of postseismic stress diffusion that he applied to the 1965 Rat Island aftershock sequence. He argued that the observed sequence was consistent with nonlinear rather than linear flow in the asthenosphere. Some objections were raised to Melosh's use of a viscous rather than a viscoelastic rheology for the asthenosphere (Savage and Prescott, 1978b; Melosh, 1978). Melosh and his co-workers subsequently developed a finite-element code TECTON to run numerical simulations with a linear or power-law viscoelastic rheology. They used this program to study the role played by surface viscoelastic deformation in determining subduction-zone topography (Melosh and Raefsky, 1980) and postseismic rebound (Melosh, 1983; Melosh and Raefsky, 1983). Figure 12 shows the postseismic uplift predicted by the Melosh dip-slip model. Notice again that when only a fraction of the elastic lithosphere is ruptured by the earthquake, the dominant characteristic of the postseismic deformation pattern is subsidence over the rupture plane. Uplift occurs in the periphery on both the overthrust and the underthrust blocks. However, when most or all of the elastic lithosphere is ruptured coseismically, the pattern changes markedly. The dominant characteristic becomes uplift centered near the surface projection of the down-dip end of the coseismic slip plane. This change in behavior is due to the fact that the

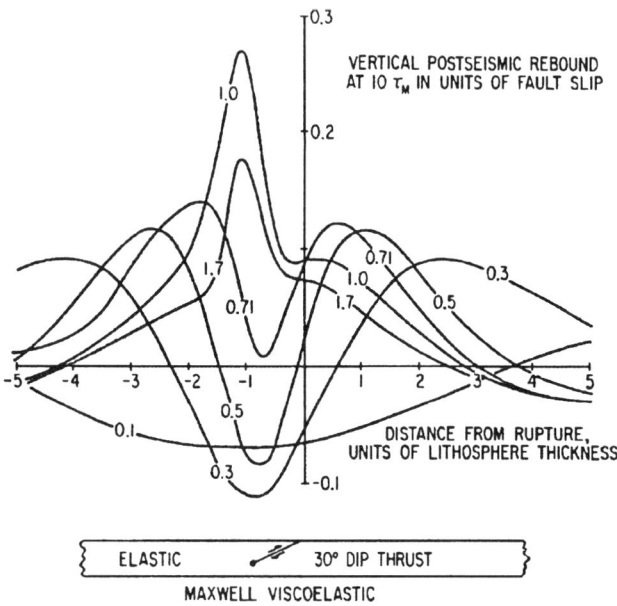

FIG. 12. Postseismic vertical motion due to viscoelastic flow in the asthenosphere following a thrust earthquake. The results are shown at ten Maxwell times and the curves are labeled by the ratio of fault depth to elastic lithosphere thickness. (From Melosh 1983.)

lithosphere undergoes coseismic shortening when the entire lithosphere is ruptured, but experiences local extension beneath the fault when only a fraction of the lithosphere is ruptured (Melosh, 1982). Thus the boundary condition on the top of the asthenosphere right after an earthquake depends on the extent to which the fault ruptures the lithosphere. When a fault with a dip angle of 30° ruptures the entire lithosphere, the maximum amplitude of the postseismic uplift ultimately reaches about 28% of the coseismic slip.

Cohen (1984) used a modified version of TECTON in a parametric study of postseismic rebound to a dip-slip earthquake using a three-layer model consisting of an elastic upper lithosphere, a standard linear solid lower lithosphere, and a Maxwell viscoelastic asthenosphere. However, one of the most interesting findings occurs for the simple case of an elastic layer over a viscoelastic half-space. In this case, the postseismic subsidence is a maximum when only one-half of the elastic lithosphere is ruptured in an earthquake. Although the coupling between the seismic slip and asthenospheric flow is greater when the slip region is extended through more of the elastic layer, the greater coupling manifests itself in a transition from subsidence being dominant to uplift being dominant. Figure 12 illustrates this point graphically, although there is a slightly discordant result in that the maximum subsidence does not appear exactly when half the lithosphere is ruptured coseismically.

Another attack on the problem of deformation due to viscoelastic flow in an asthenospheric channel underlying a lithosphere with a long, but not infinite, transform fault has been presented by Lehner *et al.* (1981). The elastic lithosphere and viscoelastic asthenosphere are coupled by a shear traction across their interface with the equilibrium equation for thickness-averaged variables being

$$\frac{\partial \sigma_{\alpha\beta}}{\partial x_\alpha} = \frac{\tau_\beta}{H}, \tag{74}$$

where H is the lithospheric thickness and τ_β is the shear traction at the interface. A tractable two-dimensional differential equation was derived by replacing the elastic stress–strain relations with the approximations

$$\sigma_{xx} = (1 + \nu)^2 \mu \frac{\partial u}{\partial x}, \qquad \sigma_{xy} = \mu \frac{\partial u}{\partial y}, \tag{75}$$

where u is the fault parallel displacement, x is the fault parallel distance, and y is the fault normal distance. The resulting differential equation is

$$\frac{\partial u}{\partial t} = \left[\alpha + \beta \frac{\partial}{\partial t}\right] \left\{(1 + \nu)^2 \frac{\partial^2 u}{\partial x^2} + \frac{\partial^2 u}{\partial y^2}\right\}, \tag{76}$$

where $\alpha = hH\mu/\eta$ and $\beta = bH$. The displacement, stress, and strain associated with these equations are averaged over the lithospheric thickness. The differential equation is solved using Fourier and Laplace transform techniques along with the Wiener–Hopf methods. Based on their results, Lehner *et al.* (1981) suggested that stress diffusion is a mechanism for the prolonged occurrence of aftershocks and for the time-delayed coupling of great earthquakes. One of the concerns associated with this model, as with others that use depth-averaged variables, is that the coupling between the elastic upper layer and the viscoelastic channel is stronger than would be the case if only a portion of the elastic lithosphere is ruptured by an earthquake.

Dmowska *et al.* (1988) used a similar mathematical model to study the stress transfer between a subduction-zone interface and the overriding and subducting plates. Their model is consistent with observations of (a) the occurrence of tensional outer rise earthquakes following large subduction-zone events, (b) compressional events occurring late in the earthquake cycle, (c) extensional intraplate earthquakes occurring down-dip from the megathrust zone late in the cycle, and (d) precursory seismic quiescence. We will review some other models which employ thin-channel flow after discussing the earthquake cycle in more detail in the next few sections.

3.3. Kinematic Models of the Entire Earthquake Cycle

3.3.1. Strike-Slip Faulting

3.3.1.1. Decomposition into steady-state and periodic terms. One of the most influential analytical analyses of the earthquake cycle using dislocation techniques has been that due to Savage and Prescott (1978a). They dealt with strike-slip motion on an infinitely long fault. As before, the fault was embedded in an elastic lithospheric layer of thickness, H, that lay over a Maxwell linear viscoelastic asthenospheric half-space. Between earthquakes, the fault is locked from the surface to depth, D, which is less than or equal to H. Consider the pattern of fault motion due to a sequence of periodic earthquakes driven by plate motion. The two plates move past one another at a constant relative velocity, $2v$, in the far-field (n.b., in Section 1.1, the relative velocity was v). At depths less than or equal to D, fault or plate-boundary slip occurs only at the time of the earthquake. As shown in Figure 13a, we can decompose the crustal motion into steady-state and cyclic terms. The steady-state term is associated with rigid motion of one plate past the other. It gives rise to translation, but no deformation. The cyclic term is driven by the sawtooth displacement-versus-time curve shown in the figure. It consists of imagined constant-velocity back slip between earthquakes and instantaneous forward slip during the earthquake. The elastic solutions to the constant-velocity back-slip and instantaneous forward-slip problems are similar to (57), but the slip function is now time-dependent. From the elastic solution, we can generate the viscoelastic solution using the Correspondence Principle described earlier. We have to consider both the slip function and the rheological parameters in using the Correspondence Principle to transform from an elastic to a viscoelastic solution. First, we replace the fault-slip function of the cyclic term (the only term which enters the calculation), $b(t)$, with is Laplacian transform. Since

$$b(t) = -2vt \quad \text{for } 0 < t < T, \tag{77}$$

$$b(t) = 0 \quad \text{for } t < 0 \text{ and } t > T, \tag{78}$$

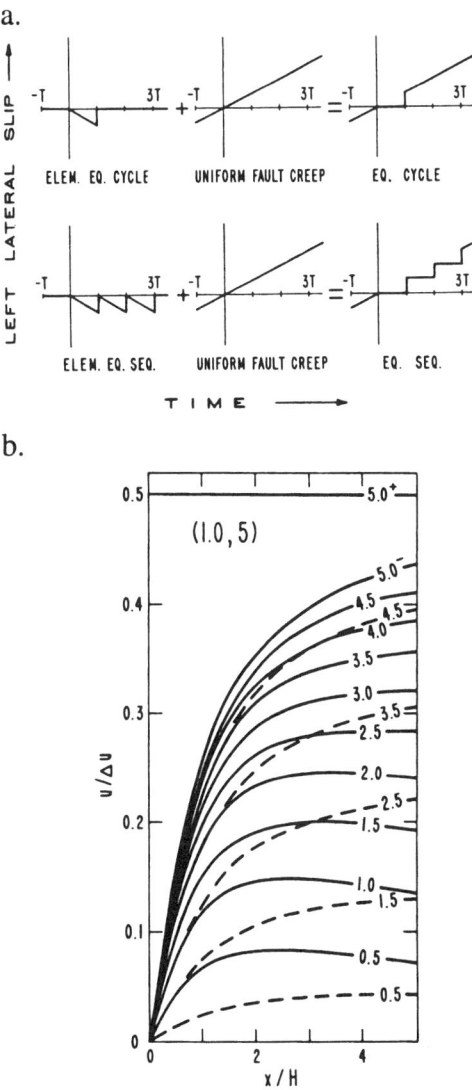

Fig. 13. (a) Diagrams showing how an earthquake sequence can be generated from superposition of a cyclic slip and uniform fault creep. The cyclic term consists of interseismic back slip and coseismic forward sliding. Uniform fault creep occurs below the locking depth, D. (b) Surface displacement versus distance from the fault at various times. In this example, $D/H = 1$ and $\tau_0 = 5$, where H is the lithosphere thickness and τ_0 is defined in the text immediately after (85). (From Savage and Prescott 1978a.)

where T is the earthquake recurrence time, the Laplace transform of the slip function is

$$\bar{b}(s) = -(2v/s^2)[1 - (sT + 1)\exp(-sT)]. \tag{79}$$

Next, we consider the terms involving the shear moduli of the lithosphere, μ_1, and the asthenosphere, μ_2. From (57), the quantity to be considered is

$$\Gamma = \left[\frac{\mu_1 - \mu_2}{\mu_1 + \mu_2}\right]. \tag{80}$$

We assume the elastic shear moduli of the lithosphere and the asthenosphere are the same ($= \mu$) but replace μ_2 by the Laplace transformed modulus of a Maxwell substance, i.e., $\mu s/(s + \mu/\eta)$, so (80) is replaced by

$$\bar{\Gamma}(s) = \left[\frac{\mu/2\eta}{s + \mu/2\eta}\right]. \tag{81}$$

The solution for a single earthquake cycle is obtained by substituting (79) and (81) into (57) and computing the inverse transformation. However, this is not the end of the calculation, for the effects of viscoelasticity may last over several earthquake cycles; thus, it is necessary to consider a sequence of N earthquakes. The general solution that takes into account multiple earthquakes cycles can be written

$$u(x,t) = \pm vt + \sum_{m=0}^{N} u_2(x, \tau - m\tau_0), \tag{82}$$

where

$$u_2 = 0 \quad \text{for } t < 0, \tag{83}$$

$$u_2 = \frac{2vT}{\pi}\left\{\left(\frac{t}{T}\right)\left[\mp\frac{\pi}{2} + \tan^{-1}(x/D)\right]\right.$$

$$\left. - \sum_{n=1}^{\infty} B_n(\tau)\tan^{-1} W_n\right\} \quad \text{for } 0 < t < T, \tag{84}$$

and

$$u_2 = \frac{2vT}{\pi}\sum_{n=1}^{\infty}[B_n(\tau - \tau_0) - B_n(\tau)$$

$$+ A_n(\tau - \tau_0)]\tan^{-1} W_n \quad \text{for } t > T. \tag{85}$$

Here $\tau = \mu t/2\eta$, $\tau_0 = \mu T/2\eta$, and $W_n = (2Dx/(4n^2H^2 - D^2 + x^2))$. The A_n are incomplete gamma functions. Also $B_1(\tau) = [\tau - A_1(\tau)]/\tau_0$, and

$B_n(\tau) = B_{n-1} - A_n(\tau)/\tau_0$. Figure 13b shows an example of the spatial and temporal dependence of the displacement pattern as predicted from the Savage and Prescott (1978a) model. Several aspects of the figure deserve particular attention. First note that, as found by Nur and Mavko (1974), the displacements in the viscoelastic solution are greater than in the elastic case. Compared to the elastic solution, the viscoelastic rates are faster in the early stages of the seismic cycle, but slower later on. This time-dependent behavior has an important implication for earthquake preparedness, for it indicates that a small strain rate can be associated with the mature stage of the seismic cycle, and that an earthquake will occur sooner than would be expected with such a small, but constant, strain rate. Note also that as deformation increases immediately after the earthquake, the displacements do not increase monotonically with distance from the fault as they do in an elastic model; rather, they reach a maximum at some intermediate distance. The location of this maximum, which also marks a change in the sign of the shear strain, diffuses away from the fault as time increases. The example shown in the figure is based on a viscoelastic relaxation time of one-tenth the recurrence time between earthquakes and a locking depth equal to the elastic lithosphere thickness. As the ratio of viscoelastic time to recurrence time increases, viscoelastic effects become less pronounced. Similarly, viscoelastic effects become less pronounced as the ratio of the thickness of the locked layer to that of the entire elastic lithosphere decreases.

3.3.1.2. Effects of viscous flow in a channel. In the Savage and Prescott (1978a) model, the viscoelastic asthenosphere extends to infinite depth. Lehner and Li (1982) addressed the question of whether the surface deformation is substantially altered when the viscoelastic flow is confined to a finite-width channel. They concluded that "a genuine stratification effect... manifests itself during the earlier part of an earthquake cycle in an amplification and concentration of post-seismic displacements" near the fault. Their model, a modification of the classical stress diffusion model of Elsasser (1969), was reexamined in a finite-element context by Cohen (1984), who found the opposite result, namely, that the early deformations were less in the channel flow model than in the half-space model, although certainly larger than in a purely elastic model. He argued that some of the depth-averaging assumptions required to develop the closed-form solutions of Lehner and Li (1982) restrict the validity of the model to the far-field, i.e., to distances greater than a few times the fault depth. Nevertheless, Li and Rice (1987) extended the channel flow model and developed a technique for extracting the surface displacements from the lithospheric averaged ones. An important aspect of their work is that the fault slip is not imposed as it is in the models of Savage and Prescott (1978a), Thatcher (1983), and Cohen and Kramer (1984), among others. The system is driven by imposing plate motion on the lithosphere at large distances from the plate boundary, allowing free slip on the fault below the seismogenic layer, and considering the coupling between the lithosphere and the asthenosphere. Li and Rice (1987) applied their model to strain accumulation on the San Andreas fault in California and estimated a locking depth of 9–11 km, an elastic layer thickness of 20–30 km, and an effective viscosity of the lower crust between 2×10^{18} and 1×10^{19} Pa s.

To investigate crustal deformation on the San Andreas fault, Ward (1985) adopted the four-layer model of Cohen (1982) consisting of an elastic upper lithosphere, a standard linear solid lower lithosphere, a Maxwell viscoelastic asthenosphere, and an elastic mantle below the asthenosphere. His model predicts that the time it takes for stress to reaccumulate following an earthquake varies with depth. An implication is that inferences of recurrence times based on surface observations may be biased.

3.3.1.3. Three-dimensional models. Although this section is devoted to three-dimensional, viscoelastic models of the earthquake cycle, it should be mentioned that computationally simple, three-dimensional, elastic models have also proven to be quite useful. Fluck *et al.* (1997), for example, have used Okada's (1985, 1992) elastic dislocation equations in a model of the Cascadia subduction zone. A feature they included in their model is a transition zone between fully locked and fully unlocked segments of the plate interface. In the preferred model, both the locked zone and the transition zone have a width of 60 km. Assuming that both zones can fail during a rupture, they concluded that the model permits earthquakes as large as $M_w = 9$.

We now turn to the main topic of this section, namely, some three-dimensional models of postseismic and interseismic deformation due to viscoelastic flow. One such model was studied by Yang and Toksoz (1981). The novel aspect of their strike-slip model was a lateral variation in viscosity with the low-viscosity material rising to shallower depths near the fault. With this geometry, the surface deformation is concentrated in the near-field. In addition, the vertical displacements, which are focused on the edges of the slip region for finite strike-slip faults, are altered by the geometry of the viscoelastic region.

Williams and Richardson (1991) developed a layered three-dimensional viscoelastic model of the San Andreas fault in central and southern California. They included the effects of nonlinear viscoelastic flow, creep at depth, and isostatic relaxation. They also allowed for surface creep as observed in central California. Several different representations of the viscosity variation with depth were considered. The starting point was a linear model with the usual elastic layer over a viscoelastic half-space; however, some of the nonlinear models allowed for a moderately ductile lower crust. The Maxwell time of the upper crust (at a strain rate of 2×10^{-14} sec^{-1}) was $\sim$ 3000 yr in most simulations, but less than 1000 yr in one case. By running a large number of finite-element simulations over a wide range of model parameters, they concluded:

(a) lowering the effective (nonlinear) viscosities in the layers near the surface results in increased deformation at all times in the earthquake cycle,

(b) lowering the effective viscosities also produces subsidence near the portion of the fault undergoing surface creep,

(c) models which include constant-rate creep at depth display less variation in surface deformation than those which have only viscoelastic flow and not steady-state creep,

(d) uplift rates are particularly sensitive to the depth of steady-state creep at bends in the fault.

3.3.2. Subduction-Zone and Other Dip-Slip Faulting Environments

One of the earliest viscoelastic models for the earthquake cycle in subduction zones was due to Bischke (1976) (see also Koseluk and Bischke, 1981). This detailed, yet often overlooked, model consisted of both shallow locked and deeper unlocked fault segments, an elastic lithosphere, a viscoelastic asthenosphere, and an elastic, or at least highly viscous, mesosphere. Several boundary conditions were investigated, including driving the oceanic plate from the side and applying tractions to the base of the lithosphere. The effects of gravitational loading were considered as well. Bischke found that he could not distinguish on the basis of available data between linear and nonlinear flow laws for the asthenosphere because the resultant surface postseismic deformation patterns were spatially similar. This conclusion was later confirmed by Melosh and Raefsky (1983). Bischke applied his model to the seismic cycle at Shikoku, Japan, site of the 1946 Nankaido earthquake. He paid particular attention to modeling the vertical motion, showing, for example, that most of the postseismic rebound takes place on the overthrust block—presumably because the largest coseismic stress changes occur at the plate interface, which lies underneath a portion of that block.

As mentioned above, Rundle and co-workers have developed an extensive mathematical apparatus for modeling crustal deformation in both dip-slip and strike-slip environments. Thatcher and Rundle (1979) consid-

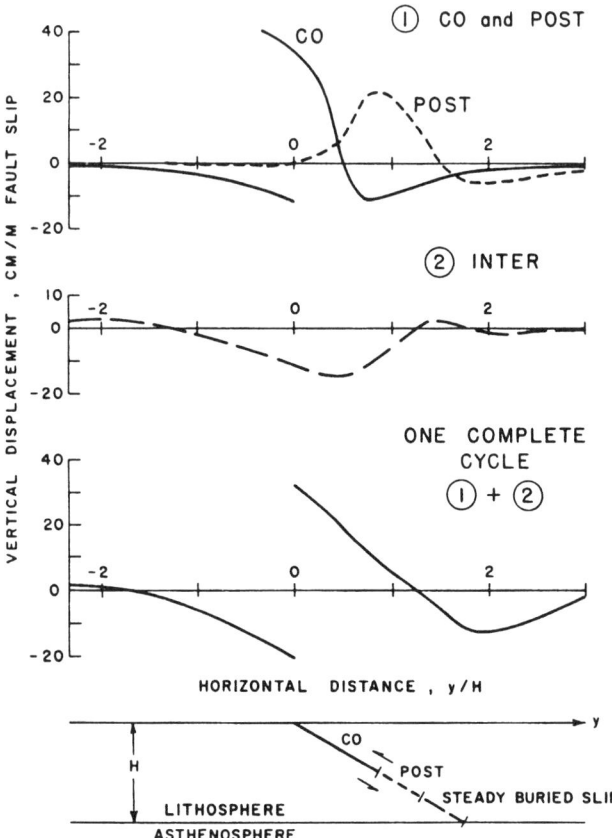

FIG. 14. Earthquake cycle mode incorporating buried steady creep and postseismic transient creep and viscoelastic effects in a thrust fault environment. The top two panels show the uplift due to coseismic, postseismic, and interseismic contributions. The postseismic and interseismic signals contain contributions from both the deep fault slip and viscoelastic rebound. The bottom panel is a sketch of the model geometry. (From Thatcher and Rundle 1979.)

ered the dip-slip model shown in Figure 14. Their model included both asthenospheric flow and fault creep at depth. The vertical motion curve labeled CO (coseismic) shows the effects of the coseismic slip on the shallow portions of the plate-boundary megathrust. POST (postseismic) has two components. The first is the influence of aseismic creep on a down-dip portion of the megathrust, and the second is the viscoelastic response to both the sudden coseismic slip and the aseismic afterslip. The term INTER (interseismic) reflects the effects of steady buried slip and the viscoelastic response to this slip; therefore, both the POST and the INTER terms contain a superposition of the effects of creep and viscoelastic flow. The viscoelastic flow produces subsidence, whereas the buried creep produces uplift at most locales, at least for the model parameters used in this calculation. It turns out that the interseismic vertical motion is dominated by viscoelastic effects. Permanent deformation accumulates through multiple earthquake cycles since the coseismic motion does not cancel the postseismic and interseismic motions. Thatcher and Rundle (1979) applied their model to earthquakes in southwest Japan, near the locales of the 1923 South Kanto and the 1946 Nankaido earthquakes. They concluded that interseismic subsidence in both regions is well explained by the asthenospheric relaxation, whereas postseismic movements are largely due to buried slip. Later, Thatcher and Rundle (1984) updated their model by taking into account gravitational (buoyancy) forces, a more complete description of steady-state plate convergence, and the cycle of repeated earthquakes, but we will defer a discussion of these results until we focus more attention on the tectonics of specific locales.

As embodied in (82) through (85), there is a mathematically simple viscoelastic model for crustal deformation in an infinitely long strike-slip environment. Unfortunately, there is no correspondingly simple solution for the dip-slip problem. Savage and Gu (1985) developed an analytical approximation for postseismic rebound, but their analysis is valid only for a single earthquake in the limit of infinite time elapsed since the event. Furthermore, the analysis is not valid close to the fault trace. Earlier, Savage (1983) had conceptually extended the viscoelastic model of Savage and Prescott (1978a) from transform to subduction-zone environments, although the ideas were not fully developed mathematically. Again, the earthquake cycle was decomposed into two terms: a steady-state subduction term that does not produce deformation, and a cyclic term, consisting of virtual, interseismic back slip along the locked portion of the plate interface and coseismic forward slip along the same interface. An assumption inherent in this model is that there is no build-up of permanent deformation in subduction zones. While this assumption is at variance with the existence of terraced beaches, it is a reasonable assumption to make in some locales, particularly over limited time spans. This model is widely employed in the literature in both its viscoelastic and its elastic forms. In the elastic case, of course, the only contribution to the interseismic deformation is that due to the virtual backward slip.

Douglass and Buffett (1995) took issue with certain aspects of the virtual slip subduction-zone models, arguing that the stress distribution that they produce on the plate interface is unrealistic, i.e., that the subducting plate exerts a traction on the overthrust plate that promotes, rather than resists, thrusting. Savage (1996) pointed out, however, that the stress which is due to the steady-state subduction cancels out this anomaly. He also pointed out that the conventional model does require significant shear stress, ~ 10 MPa, to drive the stable sliding on the plate interface immediately down-dip from the seismogenic zone.

In a series of papers, Matsu'ura and Sato (Sato and Matsu'ura, 1988; Matsu'ura and Sato, 1989, and Sato and Matsu'ura, 1993) presented an alternative to both the Savage and the Thatcher and Rundle models which differs in physical detail as well as in computational method. In particular, their work stands in contrast to that of Savage (1983) in that permanent deformation accrues. The model takes into account gravitational stresses due to the coseismic and interseismic departures from isostatic equilibrium. The implementation of the model includes changes in fault dip angle with depth, and the mathematics makes use of the hereditary integral, shown in (86) below. Let $f(t)$ be the time-dependent fault-slip function; then the surface displacement is

$$u_i(x,t) = \int_0^t d\tau \left[\frac{df}{dt}\right] \int_\Sigma q_i(x, t - \tau; x', 0) \, dx', \qquad (86)$$

where $q_i(x,t; x',0)$ is the surface displacement due to unit step slip at a point x' on the dislocation surface, Σ at $t = 0$. The functions, q_i, are the Green's functions for the viscoelastic problem. Matsu'ura and Sato (1989) discussed the explicit form for the vertical Green's function. The slip function, $f(t)$, again has two components, one due to the steady slip and one due to repeated episodes of coseismic forward slip and interseismic back slip. However, in this case, the former function does not, in general, vanish. For the steady-slip term, the slip function is $f(t) = vtH(t)$, where now H is the Heaviside function. For the cyclic term, $f(t) = -vt + vT\sum_{k=1}^{n} H(t - kT)$ for t such that $nT \leq t \leq (n + 1)T$. Referencing the displacements to positions that exist right after an earthquake, the displacement, $\Delta u(x, t)$, in an infinite sequence of earthquakes is

$$\Delta u(x,t) = vtu_s(x, \infty) - vtu_p(x, \infty) + vT \sum_{k=0}^{\infty} \left[u_p(x, kT + t) - u_p(x, kT)\right] \qquad (87)$$

for $0 \leq t < T$, and where $u_s(x,t) = \int_{\Sigma_s} q(x,t; x',0) \, dx'$ and $u_p(x,t) = \int_{\Sigma_p} q(x,t; x',0) \, dx'$. The integral for u_s is evaluated over the entire elastic plate interface, Σ_s, but the integral for u_p is integrated only over the locked portion of the interface, Σ_p. The first two terms of (87) can be grouped together, for they represent the response due to steady creep below the seismogenic layer. If the entire elastic lithosphere is ruptured in an earthquake, this term vanishes, but otherwise it increases linearly with time. When u_s and u_p are evaluated at $t = \infty$, they are known as the relaxed asthenosphere responses. The term containing the summation sign represents the viscoelastic rebound to a series of earthquakes, with the most recent earthquake being the one for which $k = 0$. For a viscoelastic rheology, this term behaves nonlinearly with time. If the entire elastic lithosphere is locked between earthquakes, then the reloading for the next earthquake comes exclusively from this rebound term. Figure 15 shows an example of the vertical deformation predicted by the Matsu'ura and Sato model. The figure shows that permanent uplift develops on the overthrust block and permanent subsidence develops on the underthrust block. A meritorious feature of this model is that it allows for the development of topography at subduction-zone boundaries; however, a weakness is that the topography will continue to grow without bound over long times. In their later work, Sato and Matsu'ura (1993) attempted to overcome this weakness by allowing for partial viscoelastic relaxation of the lithosphere.

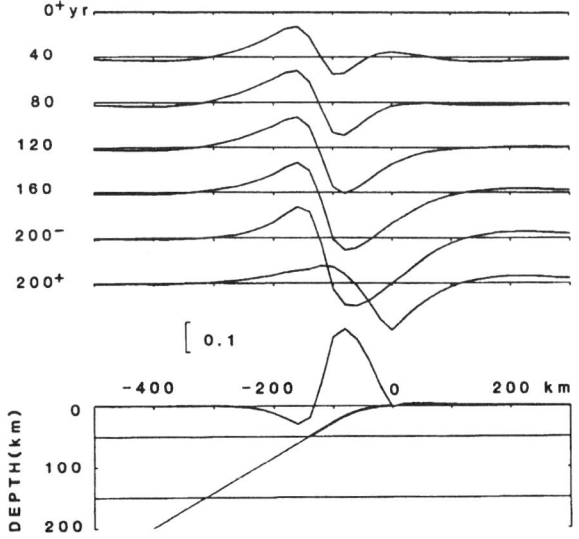

FIG. 15. Cyclic pattern of vertical displacement in subduction zones. The upper panel shows the displacement versus distance at various times during the 200-yr earthquake cycle. The lower panel shows the fault model. The locked portion of the plate interface (indicated by the heavy line) extends through the entire lithosphere. Notice that the dip angle varies with depth. (Matsu'ura and Sato, 1989; reproduced with permission from *Geophys. J. Int.*, ©1989 Blackwell Science Ltd.)

Cohen (1994) discussed the differences between the models developed by Matsu'ura and Sato and by Savage. Equation (87) can be used to highlight the issues. In the Savage (1983) model, the term involving vtu_s is assumed to be zero and the term $-vtu_p$ is the response to the virtual backward slip. As mentioned, if the entire elastic lithosphere is ruptured during an earthquake, then the relaxed asthenosphere functions vanish and the two models become conceptually identical.

The preceding discussion points out a very important and as yet unresolved question for modeling subduction-zone deformation, that is, how much permanent deformation accumulates during an earthquake cycle? More fundamentally, the question of what controls the long-term development of topography in subduction zones requires an integration of the relatively short-term (10^2–10^3 yr) models of the earthquake cycle with longer-term (10^5–10^7 yr) geological models, a task that has not yet been accomplished. Another issue which has not been fully studied for subduction-zone environments is the role played by the subducting slab. Melosh and Fleitout (1982) considered what happens when the oceanic lithosphere penetrates a viscous mesosphere underlying a fluid asthenosphere. The analysis assumes that subduction has been going on long enough for the slab to sink several hundred kilometers vertically into regions below the asthenosphere. If this is the case, the slab will act as a "sea anchor" for the subducting plate and most of the crustal deformation due to plate convergence occurs in the overthrust block. There is another factor to be considered, however, and that is the disruption of viscoelastic flow by the subducting slab. Low-viscosity material underlying the oceanic plate is shielded from the slip-generated stresses by the slab, flow cannot propagate past the plate boundary, and a mechanically strong slab partially supports the vertical stresses. In the author's opinion, these effects have not been adequately studied.

The subduction-zone models we have considered so far make an implicit assumption that all of the interseismic plate motion goes into strain accumulation. However, studies (e.g., Pacheco *et al.*, 1993) have shown that many subduction zones are weakly coupled, i.e., only a fraction of the plate motion is accounted for by seismic slip, with the rest of the convergence being accommodated by aseismic slip. As discussed by Savage and Lisowski (1986), for example, the seismic gap at the Shumagin Islands in the Aleutian arc may be a zone of weak coupling. Taylor *et al.* (1996) and Zheng *et al.* (1996) introduced a coupling factor by writing the nonseismic part of the fault-slip history as $(I - \alpha)Vt$, where each earthquake introduces a slip = αVT. They used this model with a coupling parameter, $\alpha = 0.15$, to model the uplift, tilt, and strain history of the Shumagin Islands, Alaska. While this model for the Shumagins used weak coupling,

Kozuch et al. (1996) used a fully coupled subduction-zone model for the Caribbean. Their estimated viscoelastic relaxation time of a few years corresponds to an asthenospheric viscosity of less than 3×10^{18} Pa s. The authors observed that their model predicts that the current epic is about 45–60 years into the earthquake cycle, but the last great subduction-zone earthquake in the region occurred over 100 years ago.

As the variability in plate coupling implies, the resistance to motion that occurs at plate interfaces is not necessarily spatially uniform. Patches of the plate interface that are highly resistant to slip, called asperities, concentrate deformation and are the loci of high moment release in an earthquake. These asperities can sometimes be identified with specific geological structures such as subducting seamounts or variations in subduction geometry. Dmowska et al. (1996) studied the effects of asperities on seismicity and deformation. They noted that seismically active areas of an outer rise and slab are sometimes offset along strike from the asperity locations in obliquely convergent zones. Using the model shown in Figure 16, they found that slip obliquity has a significant effect on the deformation pattern and can produce the observed offset in the location of seismically active zones with respect to the asperity.

3.4. Other Issues

3.4.1. Spherical Models

Dragoni et al. (1983) and Sabadini et al. (1984) considered viscoelastic rebound for a spherically stratified, incompressible earth. The approach adopted in the later paper was to solve the coupled conservation of momentum and Poisson's equations, subject to the incompressibility constraint. The conservation of momentum equation is

$$\nabla \cdot \tau - \rho_0(r)\nabla\Phi_1 - \rho_1 g_0(r)\hat{r} - \nabla(u \cdot \rho_0 g_0 \hat{r}) + f = 0, \qquad (88)$$

where ρ_0 is the density, $g_0(r)$ is the gravitational acceleration, $\hat{r}$ is the radial unit vector from the center of the Earth, and f is any static body force applied to the system. The density perturbation, ρ_1, is obtained from the incompressibility constraint,

$$u \cdot \left(\frac{\partial \rho_0}{\partial r}\right)\hat{r} + \rho_1 = 0. \qquad (89)$$

Poisson's equation for the gravitational potential, Φ_1, is

$$\nabla^2 \Phi_1 = 4\pi G \rho_1, \qquad (90)$$

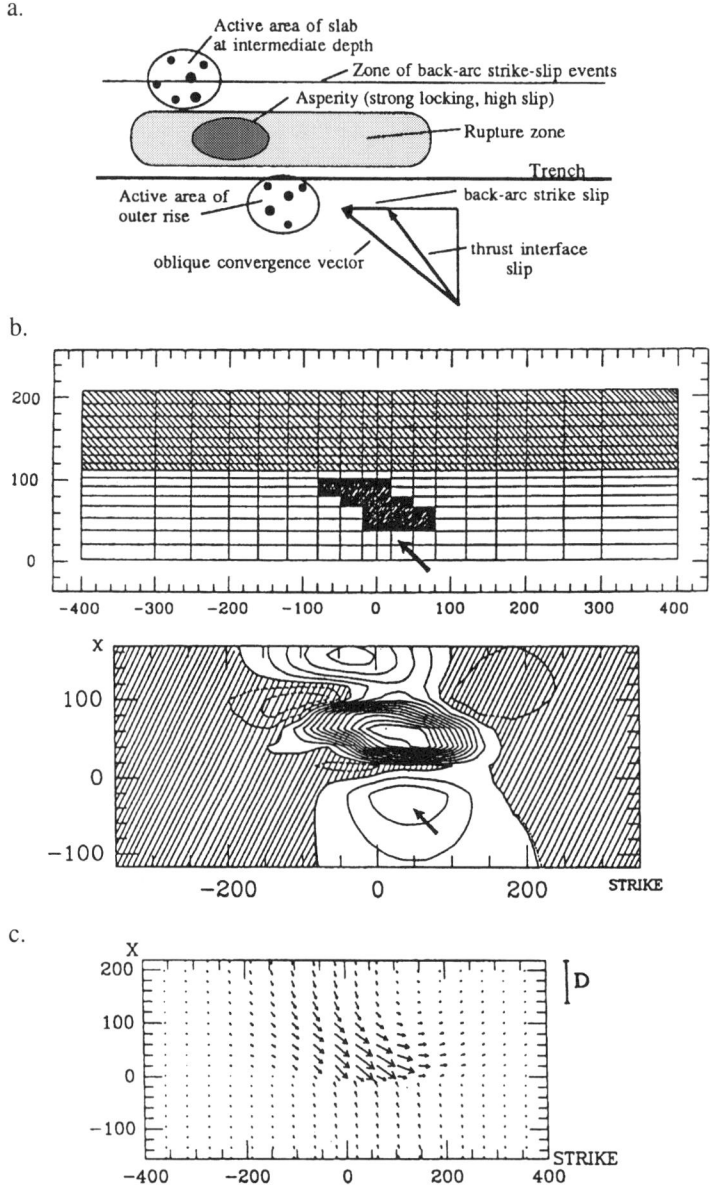

FIG. 16. Stress and deformation associated with an asperity located at an obliquely convergent plate boundary. (a) Cartoon shows schematically how the location of seismically active portions of subducting slab and outer rise can be offset along strike from the asperity. (b) Finite-element model of asperity (dark patch) and computed horizontal stress for oblique slip, D, at 45°. (c) Horizontal displacements. (From Dmowska et al. 1996.)

where G is the universal gravitational constant. The displacement field is expressed in terms of scalar potentials, U and V, by writing

$$\boldsymbol{u}(\boldsymbol{r}) = U(r, \theta, \phi)\hat{r} + \frac{\partial}{\partial \theta}V(r, \theta, \phi)\hat{\theta} + \frac{1}{\sin \theta}\frac{\partial}{\partial \phi}V(r, \theta, \phi)\hat{\phi}. \quad (91)$$

The displacement and gravitational potentials are then expanded in terms of spherical harmonics, $Y_l^m(\theta, \phi)$, by writing, for example,

$$U(r, \theta, \phi) = \sum_{l=0}^{\infty} \sum_{m=-l}^{l} U_l^m(r) Y_l^m(\theta, \phi). \quad (92)$$

The resulting differential equations are solved by numerical techniques.

One major research thrust which motivated the development of spherical models of crustal deformation was the assessment of the effects of faulting on excitation of the Chandler Wobble. As we have already discussed, it turns out that the excitation due to the coseismic deformation is small. While inclusion of the effects of viscoelasticity can enhance changes in the inertia tensor by a factor of 4 or more over a longer term, the excitation of the Chandler Wobble remains small because the time scale for the inertia changes is not appropriate for wobble excitation. Only for very low viscosities (10^{16} Pa s) would the effects of viscoelasticity be pronounced.

In recent years, more attention has been focused directly on the geodetic implications of global crustal deformation patterns. Piersanti *et al.* (1995) considered the problem of postseismic rebound in a radially stratified and self-gravitating viscoelastic earth. For their models, the lithosphere was elastic to a depth of 100 km. In some cases the underlying mantle was assumed to have a uniform viscosity down to the mantle–core boundary, whereas in others the viscosity increased by a factor of 30 at the 670-km-depth seismic discontinuity. An earthquake was modeled as a point source, although finite sources were considered later by Piersanti *et al.* (1997). An important aspect of this work has been the comparison of the postseismic rebound theory with the very similar postglacial rebound theory. In Laplace transform space, the displacement, $\mathbf{u}$, can be separated into two terms according to $\mathbf{u} = \mathbf{u}_T + \mathbf{u}_P = \nabla \times [\overline{T}(r)e_r] + \nabla \times \nabla \times [\overline{S}(r)e_r]$, where the Laplace transformed toroidal component, $\overline{T}$, and poloidal component, $\overline{S}$, are both expanded in spherical harmonics. Piersanti *et al.* (1997) studied the normal model expansion for the displacement field and pointed out that within the context of radially stratified models, postglacial deformation contains no toroidal components. On the other hand, in postseismic rebound both toroidal and poloidal components exist.

One of the important aspects of the spherical models is their ability to predict deformations on the surface of the Earth at large distances from the source. Piersanti *et al.* (1995) carried the spherical harmonic expansion to degree 1000 and found that horizontal deformation propagates to greater distances from the source than does vertical deformation. In a later paper, Piersanti *et al.* (1997) examined geodetic data in the form of Very Long Baseline Interferometry (VLBI) baselines between a site near Fairbanks in central Alaska and several sites in Canada and central, southern, and eastern portions of the United States. They argued that continuing postseismic rebound from the aforementioned 1964 Prince William Sound earthquake in southern Alaska would affect several of these baselines at the level of about 1 mm/yr, within the detectability of the observations. This estimate was based on a dip angle of 20° for the coseismic rupture plane. However, seismic tomography work (Zhao *et al.*, 1995) indicates that the dip angle of the locked Pacific−North American interface is significantly less, so the estimates of the viscoelastic rebound effects may be biased.

As the spherical models have become more sophisticated, attempts have been made to include considerable vertical stratification. Sabadini and Vermeesen (1997), for example, used a complete 111-layer Preliminary Reference Earth Model (PREM, Dziewonksi and Anderson, 1981) representation for the elastic moduli and density structure, but they discovered that only about ten layers are resolvable.

One of the reasons for assuming incompressibility in the spherical models of crustal deformation is to simplify the integrations required to determine the inverse Laplace transforms that arise in the use of the Correspondence Principle. This assumption is suspect, however, because elastic volume changes can occur in an earthquake. Ma and Kusznir (1992) showed several theoretical examples where the subsurface volumetric coseismic strain change is about the same size as the shear strain change. Pollitz (1997) has discussed this issue and offered an alternative attack on the problem in which compressibility is included in the calculation, but certain perturbation terms in the equilibrium equation that involve the Earth's gravitational constant, G, are ignored. In addition, he assumed that within each layer the product of density and gravitational acceleration varies inversely with radius. He found that thin- and thick-asthenosphere models behave somewhat differently. As shown in Figure 17, the postseismic horizontal displacements at $t = 5\tau$ are attenuated in the thin-asthenosphere case. However, by $t = 45\tau$ the far-field horizontal displacements within the underthrust plate are somewhat larger than in the thick-asthenosphere case.

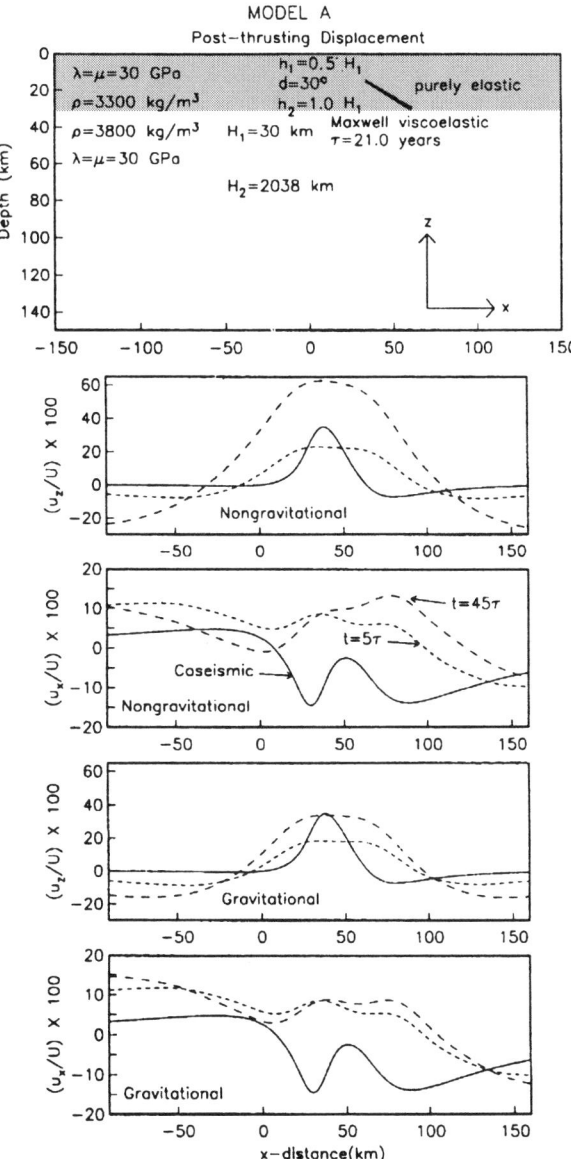

FIG. 17. Coseismic displacements (solid line) and postseismic displacement (dashed lines) following a thrust earthquake using a spherical viscoelastic model. Model A has a thick asthenosphere, model B has a thin asthenosphere. In each column the top panel shows the structural model and model parameters; the four lower panels show (from top to bottom), vertical motion—viscoelastic effects only, no gravity; horizontal motion, viscoelastic effects only, no gravity; vertical motion, viscoelastic and gravitational effects; horizontal motion, viscoelastic and gravitational effects. (From Pollitz 1997.)

194 STEVEN C. COHEN

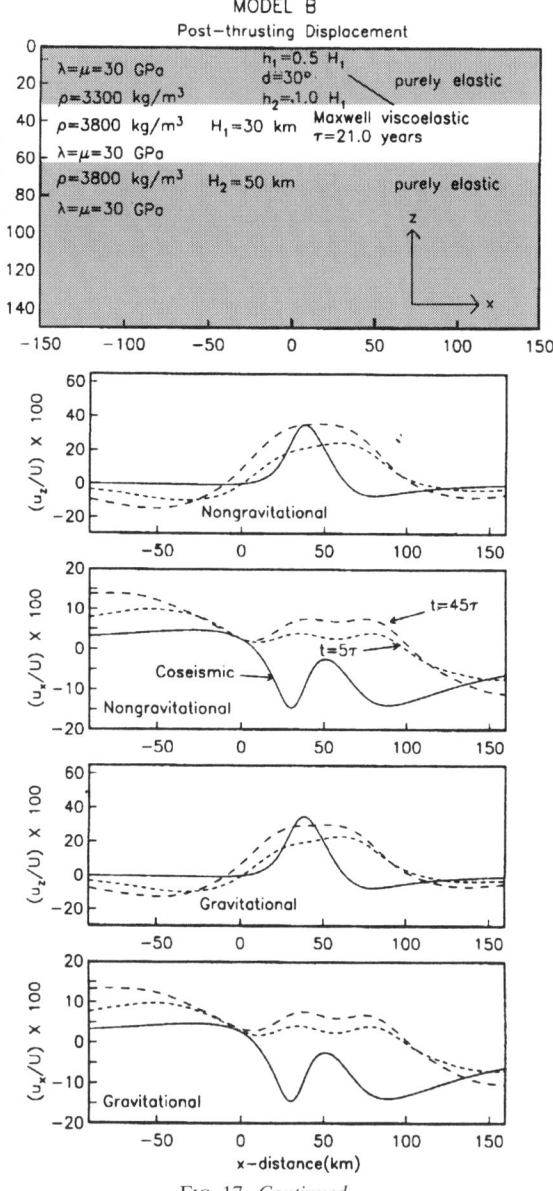

FIG. 17. *Continued.*

3.4.2. Nonlinear Flow Effects

Lyzenga *et al.* (1991) modeled motion on the San Andreas fault using a nonlinear, depth-dependent viscoelastic finite-element model in which the system was driven by imposing plate motions on boundaries distant from the fault and allowing the fault to slip when a prescribed stress accumulated on the fault. Their interest was primarily in the stress distribution within the crust. They found, for example, that there is a stress minimum at depths less than the seismogenic depth and a stress maximum at somewhat greater depths. The stress maximum is maintained by the repeated transfer of stress from the elastic crust to the ductile asthenosphere. Reches *et al.* (1994) also modeled periodic earthquakes on the San Andreas fault using a power-law rheology. They drove the model by imposing velocities of ± 17.5 mm/yr at the base and side of a 25-km-thick, 500-km-wide crustal block and let the fault slip by 2.8 m every 160 yr. They found they could best fit the geodetic observations by using a rheology that has a viscosity considerably less than that measured in the laboratory for quartzite. As a result, they concluded that the *in situ* viscosity of lower crustal rock is less than that indicated by the experiments. Of course, this conclusion assumes that viscous flow is the most important mechanism for the time-dependent deformations that are observed geodetically on the San Andreas fault, and this assumption is still without compelling empirical support.

3.4.3. Coupled Earthquakes

The occurrence of an earthquake can, in principle, stimulate or impede the occurrence of another event on the same fault or even on a remote one. One potential consequence of repeated coupling of this nature is a spatial propagation of earthquakes. Unambiguous evidence for the occurrence of propagating earthquake epicenters is rather rare, but illustrative cases can be found in north Anatolia, along the San Jacinto fault in California, and elsewhere. Among the mechanisms that have been proposed to be responsible for this phenomenon are elastic and viscoelastic transfer of quasi-static stress fields, dynamic effects associated with the passage of seismic waves, and reorganization of a critical system in which long-range interactions in a mathematically nonlinear system "are not through direct transmission of stress but through the interactions of faults with each other" (Turcotte, 1992). Dislocation theory has been used to study the most obvious of these mechanisms, the rearrangement of the local and regional scale stress field due to the occurrence of a major earthquake. Some of the more recent studies of earthquake triggering have used dislocation theory to compute changes in the Coulomb stresses on a fault (Stein et al., 1992; Hodgkinson et al. 1996). The Coulomb criterion for the occurrence of an earthquake is that failure occurs when the shear stress on the fault plane, τ, reaches the value $\tau_0 + \mu(\sigma_N - P)$, where τ_0 is the intrinsic material strength, μ is the coefficient of friction, σ_N is the normal stress on the plane, and P is the fluid pressure. The occurrence of a nearby earthquake changes the Coulomb stress, $\tau_c = \tau - [\tau_0 + \mu(\sigma_N - P)]$, with the sign of the change depending on the orientation of a potential future rupture plane relative to the surface that ruptured in the earthquake. King et al. (1997) have found that four moderate-magnitude earthquakes in southern California between 1975 and 1992 increased the Coulomb stress at the site of the 1992 $M_w = 7.3$ Landers earthquake by about 0.1 MPa (1 bar). They deduced that the time of occurrence of the Landers earthquake was advanced by one to three centuries by these four events. The Landers earthquake increased the Coulomb stress by 0.3 MPa at the site of the subsequent $M_L = 6.5$ Big Bear aftershock. While it is possible to calculate which regions have become more stressed as a result of an earthquake, these calculations do not indicate whether a subsequent earthquake will be triggered. King et al. (1997) speculated that stress changes on the order of 0.05 MPa are sufficient to trigger or suppress the occurrence of an earthquake. However, even if it were possible to know the total Coulomb stress, rather than just changes in it, the Coulomb criterion is not a fully adequate friction law for determining whether unstable (earthquake) sliding will occur. In the next section, we will discuss a more sophisticated fault friction law that can be used to study the slip stability itself.

Rydelek and Sacks (1988, 1990) have noticed a correlation between the occurrence of onshore earthquakes in Japan and offshore, subduction-zone events; the land events precede those in the subduction zone by about 36 years. They argued that the convergence of the Pacific and Asian plates causes buckling and the occurrence of earthquakes in the overthrust block. A stress change due to the onshore earthquake subsequently diffuses viscoelastically toward the subduction zone, where it modifies both the shear and the normal stress at the plate interface. This stressing is proposed as a stimulating mechanism for the occurrence of subduction-zone events. The stress changes at the plate interface are rather small; in fact, the finite-element calculations of Cohen (1992) indicate that they are even smaller than those predicted by the modified Elsasser model used by Rydelek and Sacks (1988, 1990). Nevertheless, if the model is otherwise valid, the appropriate value for the viscosity of the asthenosphere is about 10^{19} Pa s, in accord with other viscosity determinations near Japan.

3.4.4. Time, Slip, and Velocity-Dependent Friction

Most of the models we have considered up to now are kinematic in that the fault slip is specified rather than derived from basic dynamical principles. There are, however, some models which attempt to compute the time-dependent fault slip and stress. For example, it is possible to drive the fault system by specifying the velocity at distant points and allowing stress to accumulate on the fault. Slip occurs when the shear stress exceeds the static friction. In the simple case of a block sliding on a frictional surface, the amount of slip can be related to the difference between the static and the dynamic coefficient of friction. A more sophisticated model of slip has been developed by Tse and Rice (1986), who based their analysis on a fault friction law which depends on normal stress, temperature, slip rate, and slip history and, hence, is depth-dependent. The coefficient of friction associated with sliding at a constant velocity approaches a steady-state value, μ_{ss},

$$\mu_{ss}(V) = \mu_{ss}(V^*) + (a - b)\ln(V/V^*), \tag{93}$$

where V is the sliding velocity and V^* is a reference velocity for which the coefficient of friction is μ_{ss}. As cited by Lorenzetti and Tullis (1989), for an e-fold change in sliding velocity, the constitutive parameter "a" measures the magnitude of the instantaneous change in the coefficient of friction while the parameter "b" measures the magnitude of the change in the coefficient of friction which occurs during slip at the new velocity. If $a - b > 0$, then friction increases with increases in the sliding velocity and stable sliding occurs. However, if $a - b < 0$, then an increase in sliding velocity decreases the coefficient of friction. The material weakens and sliding can be either stable or unstable; thus, a system obeying this friction law can exhibit many of the characteristics of fault motion. Equation (93) can be recast in terms of steady-state shear stress, τ_{ss}, by multiplying the coefficient of friction by the normal stress, σ_n, to get

$$\tau_{ss}(V) = \tau_{ss}^* + (a - b)\ln(V/V^*)\sigma_n. \tag{94}$$

An equation for the dynamics of the system can be developing by defining a characteristic distance, L (D_c is also used), such that the frictional stress τ obeys

$$\frac{d\tau}{dt} = \frac{1}{L}[\tau - \tau_{ss}]. \tag{95}$$

Tse and Rice (1986) used (94) and (95) to compute fault displacement versus depth at various times during an earthquake cycle for a strike-slip fault. The results from a typical simulation are shown in Figure 18. Locations near the surface move very little except during the unstable sliding that characterizes an earthquake. By contrast, points at depth move continuously with time and do not experience instability sliding. At intermediate depths, there is a period of accelerated postseismic creep following the surface instability sliding. Lorenzetti and Tullis (1989) used this model to study whether short-term geodetic precursors are likely to occur with a magnitude and temporal scale large to be big enough to be observed. They used characteristic distances ranging from 5 mm to 40 mm and concluded that for all but the smallest decay distances, observable geodetic signals should occur within 1–2 km of a strike-slip fault for a few minutes to a month before a major earthquake. The fact that only a few precursory geodetic signals have been observed suggests that the decay distance may be even smaller than assumed by Lorenzetti and Tullis; however, such models are meritorious in providing for preseismic increases in the deformation rate based on fundamental physics and observed rock mechanics.

Stuart (1988) used a similar friction law to model earthquake occurrence in the Nankai Trough subduction zone. He chose numerical values for the location of the minimum value of $a - b$ and the characteristic length, L, by constraining a reference model to give the observed 92-yr recurrence time between the 1854 Ansei and the 1946 Nankaido earthquakes and also

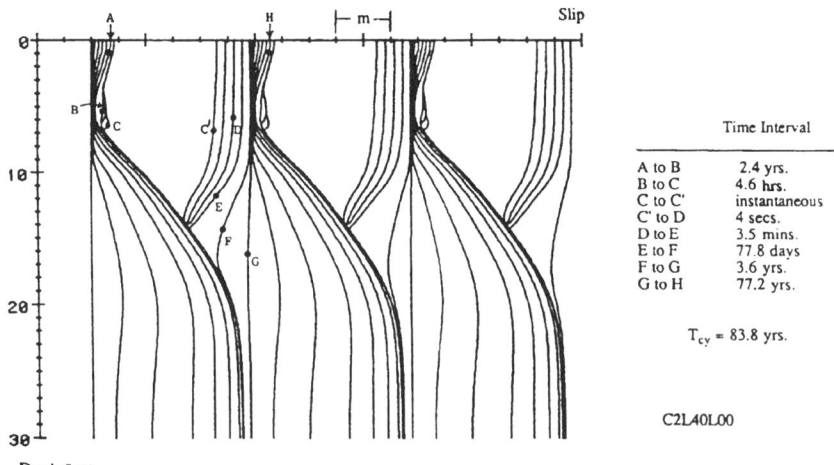

FIG. 18. Slip versus depth through a time history of three earthquakes. The stress law, (93), becomes velocity dependent when the velocity exceeds 0.1 mm/s. The slip weakening distance, L, is 40 mm. (From Tse and Rice 1986.)

to give the observed coseismic uplift of the 1946 event. He predicted that there are two time scales for the occurrence of accelerated fault slip prior to an earthquake. The first slip occurs over several years before the earthquake and involves slip over the entire brittle region. The second slip is a few days long and propagates away from a nucleation site.

3.4.5. Erosion, Deposition, and Accretion

As stated earlier, one of the greatest challenges to the development of models of crustal deformation is to relate the deformations that occur on the time scale of the earthquake cycle to those which occur on geological time scales. Several factors which are often ignored in geodetic and seismological models become quite important to the growth of geological structures. Foremost among these are the effects of erosion and deposition, including not only the direct consequences of mass movement but also the isostatic response to these movements. To date, no model has incorporated the effects of the earthquake cycle, erosion and deposition, gravitational loading, and crustal accretion in a totally self-consistent manner; however, in companion papers, King *et al.* (1988) and Stein *et al.* (1988) explored some of the mechanisms by which repeated earthquakes generate geological structures. The concepts behind their model are shown in Figure 19. A simple two-step strategy was adopted for modeling the deformation. First, the earthquake cycle deformations were obtained. The coseismic displacements were computed from dislocation theory and the postseismic and interseismic deformations were estimated by calculating the deformation field for a fully relaxed viscoelastic asthenosphere. Next, erosion (deposition) was assumed to occur in places that experienced uplift (subsidence). The deposition is a positive surface load on the system, whereas the erosion is a negative load. The crustal flexure due to these imposed loads is then added to the deformation pattern. For example, for a line load of magnitude, F, the vertical crustal flexure is

$$w = e^{-kx}\frac{Fk}{2\Delta\rho g}(\cos|kx| + \sin|kx|), \tag{96}$$

where $k^4 = 4D/(\Delta\rho g)$ and $D = EH^3/[12(1-\nu^2)]$. In these equations, E is Young's modulus, ν is Poisson's ratio, H is the thickness of the elastic plate, g is the gravitational acceleration, $\Delta\rho$ is the density difference between the viscoelastic fluid below the elastic plate and air above it, and D is the crustal flexural rigidity. The corresponding equation for a point

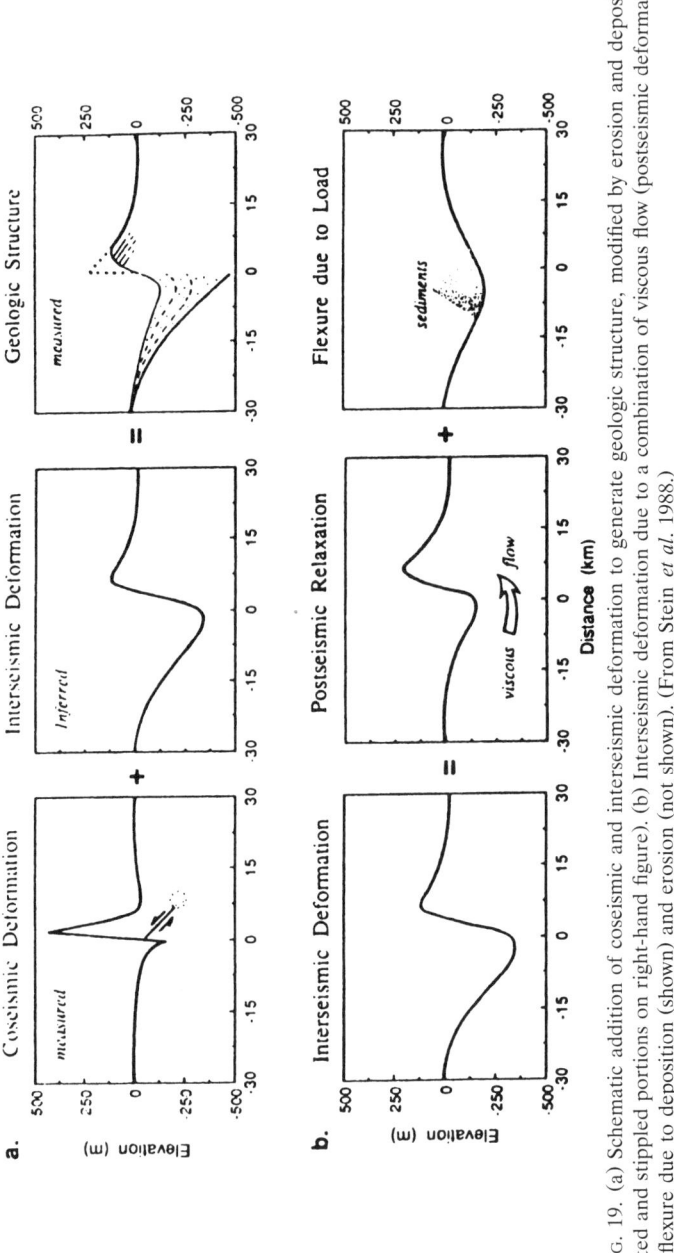

FIG. 19. (a) Schematic addition of coseismic and interseismic deformation to generate geologic structure, modified by erosion and deposition (dotted and stippled portions on right-hand figure). (b) Interseismic deformation due to a combination of viscous flow (postseismic deformation) and flexure due to deposition (shown) and erosion (not shown). (From Stein et al. 1988.)

load is

$$w = -\frac{Fl}{2\pi D}kei\left(\frac{r}{l}\right), \qquad (97)$$

where $kei(x)$ is the zero-order Kelvin function (Abramowitz and Stegun, 1972) and $l^4 = D/\Delta\rho g$. For deposition, the surface load can be estimated from seismic and geologic constraints on the sediment accumulated in regions of subsidence. For erosion, the surface load can be estimated from geological reconstructions of uplift structures. Stein et al. (1988) applied this general model to specific regions in California, Utah, and Idaho. They deduced a flexural rigidity of $2-15 \times 10^{19}$ N m^{-2}, corresponding to an elastic plate thickness of 2–4 km. Since the deduced crustal thickness is quite thin and less than the depth of faulting, the model results imply that the plate must have less strength on geological time scales than on seismological ones. This thinning of the elastic plate causes fault-bounded basins to narrow over time.

Sato and Matsu'ura (1993) incorporated the effects of accretion and deposition into a kinematic model of the evolution of island arc-systems. Although their model is based on a steady-state situation and, therefore, is not directly applicable to the earthquake cycle, it is of interest here because it is based on the steady-state behavior of their earlier model of earthquake deformation. Accretion is modeled by assuming that the removal of sediments from the top of the subduction oceanic plate and their emplacement on the overriding continental plate can be approximated by allowing the plate boundary to propagate seaward at a constant rate. Taking a coordinate system fixed on the continental plate and referring to the hereditary integral, (86), the surface vertical displacement, w_s, due to steady convergence at rate, v_p, but without accretion is

$$w_s(x,t) = v_p \int_0^t u_s(x, t - \tau) \, d\tau \quad \text{for } x_2 \leq 0 \qquad (98)$$

and

$$w_s(x,t) = v_p \int_0^t u_s\bigl(x + v_p(t - \tau)e_2, t - \tau\bigr) \, d\tau \quad \text{for } x_2 > 0, \qquad (99)$$

where x_2, distance normal to the plate boundary, is positive in the oceanic plate and negative in the continental plate, e_2 is the unit vector along the 2 axis, and, as before, $u_s(x, t - \tau) = \int_\Sigma q_s(x, t; x', \tau) \, dx'$, where q_s is the viscoelastic vertical component of surface displacement at a point, x, and a time, t, due to a unit slip at point, x', at time, τ. The effect of accretion at a rate, v_a, is to replace (98) and (99) by

$$w_s(x,t) = v_p \int_0^t u_s(x - v_a e_2 \tau, t - \tau) \, d\tau \quad \text{for } x_2 \le 0, \quad (100)$$

$$w_s(x,t) = v_p \int_0^{t_1} u_s\big(x + v_p(t_1 - \tau)e_2 - v_a \tau e_2, t - \tau\big) \, d\tau$$
$$+ v_p \int_0^t u_s(x - v_a e_2 \tau, t - \tau) \, d\tau \quad \text{for } 0 < x_2 \le v_a t, \quad (101)$$

and

$$w_s(x,t) = v_p \int_0^t u_s\big(x + v_p(t - \tau)e_2 - v_a \tau e_2, t - \tau\big) \, d\tau \quad \text{for } v_a t < x_2, \quad (102)$$

where $t_1 = x_2/v_a$ and the region $0 < x_2 \le v_a t$ is the accreted material. To model the effects of deposition (and erosion), Sato and Matsu'ura assumed that the rate of sedimentation is known and calculated the viscoelastic response to the imposition of this surface load. A final feature that is incorporated into the model is to assume that the lithosphere undergoes viscoelastic relaxation on a time scale long compared to that of the asthenosphere. This model feature imposes a cap on the build-up of topography. Sato and Matsu'ura found that their model predicts the steady uplift of marine terraces at subduction zones and stable patterns of topography and gravity anomalies after several million years of subduction. An effect of accretion is to cause these steady features to migrate seaward, thus allowing marine terraces to undergo steady uplift even at old subduction zones.

3.4.6. *Other Miscellaneous Processes*

Changes in the pore pressure in crustal rocks induced by an earthquake or other causes result in a change in the volumetric strain and, hence, can be a source of deformation. The poroelastic stress–strain relationships are (e.g., Jaeger and Cook, 1976)

$$E\varepsilon_x = \sigma_x - \nu(\sigma_y + \sigma_z) - \frac{EP}{3H} \quad (103)$$

and

$$E\varepsilon_{xy} = (1 + \nu)\sigma_{xy}, \quad (104)$$

where E is Young's modulus, ν is Poisson's ratio, P is the fluid pressure, and H is a constant whose physical significance will soon be manifest. The equation for shear is unaffected by fluid pressure, while the normal equation reduces to the usual elastic relationship when the fluid pressure vanishes. Adding the equation for ε_x to similar equations for ε_y and ε_z defining the volumetic strain, Δ, by $\Delta = \varepsilon_x + \varepsilon_y + \varepsilon_y$, using the relationship between the bulk modulus, K, and E and ν, i.e., $K = E/[3(1-2\nu)]$, we find,

$$\Delta = \frac{\sigma_x + \sigma_y + \sigma_z}{3K} - \frac{P}{H}. \qquad (105)$$

For hydrostatic conditions, when the hydrostatic pressure equals the pore pressure,

$$\Delta = P\left[\frac{1}{K} - \frac{1}{H}\right]. \qquad (106)$$

Thus, $1/H$ has the effect of reducing the volumetic compressibility. Poroelasticity is fundamentally different from viscoelasticity in that it changes the volumetric rather than the shear properties of the rock. Westerhaus et al. (1997) have suggested that strain anomalies that are observed along the North Anatolian fault are correlated with changes in groundwater flow and might, therefore, be due to pore fluid effects.

Hofton et al. (1995) extended the viscoelastic-gravitational postseismic rebound theory of Rundle to calculate the deformation due to the dilatation associated with the emplacement of a dike in the elastic layer. They initially solved the layered elastic problem using source functions published by Ben-Menahem and Singh (1968) and then invoked the Correspondence Principle to calculate the viscoelastic response when the elastic layer is underlain by a viscoelastic half-space. Vertical dikes that rupture the entire lithosphere and those that are deeply buried in the lithosphere both produce broad-scale horizontal deformation, but the amplitude of deformation is larger in the former case. Since the Lamé parameter, λ, is allowed to relax viscoelastically, the bulk modulus also relaxes, whereas rock mechanics results suggest that the bulk modulus undergoes little if any viscous modification. While the relaxation of the bulk modulus is only a minor concern for faulting problems that primarily involve shear, it may be more significant for processes which involve significant volumetric changes such as dike formation.

Bonafede (1990) studied the axisymmetric deformation of a thermo-poro-elastic half-space. His application was the deformation associated with the flow of lava into a shallow magma chamber. The importance of this paper in the present context is that it corrects some errors in the use of Maruyama's strain nuclei (Maruyama, 1964) by Bonafede et al. (1986).

4. Crustal Deformation near Specific Faults

While the preceding sections have focused on the theoretical aspects of numerical modeling, several of the results have been applied to specific locales. In this section, we take a closer look at the geodesy and seismotectonics of three locations, the Nankai subduction zone in Japan, the eastern portion of the Aleutian subduction zone in southcentral Alaska, and the San Andreas transform system in California. The Nankai subduction zone, the boundary between the subducting Philippine Sea plate and the overriding Eurasian Plate, has played a prominent role in the development of crustal deformation models. The historical record of earthquakes extends back to the seventh century and geodetic observations date from 1895 (Thatcher and Rundle, 1979), giving this region one of the most temporally complete records of crustal deformation. In the last 150 years, major earthquakes occurred in 1854, 1944, and 1946. The historical record in southern Alaska is somewhat shorter; however, the great 1964 Prince William Sound earthquake occurred at a time when numerical models of crustal deformations were first being developed in detail. Thus, the wide variety of geodetic, seismological, and geological data that were collected provided critical constraints for testing early models. The postseismic crustal deformations that followed this earthquake are among the largest recorded anywhere. The San Andreas fault system has played as much of a role in the development of models of crustal deformation for strike-slip environments as the Nankai and eastern Aleutian regions have for subduction environments. Great earthquakes in 1857 in southern California and in 1906 in north-central California were the stimulus to study the San Andreas system. Today, the San Andreas system is the most widely studied strike-slip fault system in the world.

4.1. Nankai Subduction Zone, Japan

The long history of geodetic observations in Japan has been exploited to develop some of the most detailed models of the earthquake cycle, particularly for the Nankai subduction zone, site of the 1944 Tonankai ($M = 8$) and 1946 Nankaido ($M = 8.2$) earthquakes. Fitch and Scholz (1971) were among the first to apply dislocation theory to the 1946 Nankaido earthquake. By comparing geodetic data to model predictions, they deduced that the earthquake was predominantly a dip-slip event. Their preferred fault model had a fault dip angle of 30° to 40° and involved variable slip ranging from 5 m to 18 m. They explained the rapidly decaying (time scale of a few years) postseismic deformation in terms of a combination of forward slip on the downward extension of the coseismic fault surface and backward slip on the coseismic rupture surface itself. Scholz and Kato (1978) extended the work in this region by considering the interseismic deformation for 24 years following the earthquake. Their model consisted of a locked upper portion of the plate interface and a lower slipping segment. The model which best fit observed leveling data ascribed only 20–50% of the relative plate motion to the slipping portion of the plate interface, suggesting that only a portion of the plate motion gets coupled into elastic strain accumulation and ultimately seismic slip.

As already mentioned, Thatcher and Rundle (1979, 1984) attributed early postseismic rebound in the South Kanto and Nankaido regions of Japan to creep at depth, but they claimed that most of the longer-time-scale interseismic strain accumulation was driven by viscoelastic flow. Their coseismic model uses fault parameters for the 1946 Nankaido earthquake published by Ando (1975). The viscoelastic-gravitational coupling consists, as usual, of an elastic plate over a viscoelastic half-space, but is also allows for buoyancy forces and multiple earthquake cycles. They found that the initial postseismic response to the 1946 Nankaido earthquake, as constrained by tide gauge observations and shown in Figure 20a, is better explained by creep at depth than by viscoelastic flow. Conversely, leveling data from the Kii Peninsula spanning two decades (Figure 20b) was explained by viscoelastic rebound, where the viscoelastic model assumes that the entire elastic lithosphere is ruptured in the earthquake, that the earthquake recurrence time is 100 yr, and that the Maxwell time of the asthenosphere is 2.5 yr. On yet a longer time scale, the viscoelastic model underpredicts the observed uplift unless the ratio of recurrence time to Maxwell time is increased to about 100.

Miyashita (1987) has emphasized lateral heterogeneity in the mechanical properties of the Earth due to the presence of a subducting slab. To model the plate interface between the Philippine Sea and Eurasian plates, he assumed strong coupling at shallow depths and weak coupling deeper. Slip during the 1946 Nankaido earthquake is along a thrust fault which intersects the plate boundary at depth, but has a surface exposure on the overthrust block. The region of weak coupling is represented by a thin low-viscosity layer; viscoelastic flow also occurs in the asthenosphere. A schematic representation of the model and a comparison between observed and computed preseismic, coseismic, and postseismic uplifts are shown in Figure 21. For the theoretical calculations, coseismic slip was imposed on a fault that extends from the surface to a depth of 32 km. The dip angle is about 20° at the lower depths but increases gradually toward the surface. The fault is coincident with the Asian–Philippine Sea plate boundary at depths below about 22 km, but intersects the surface west of

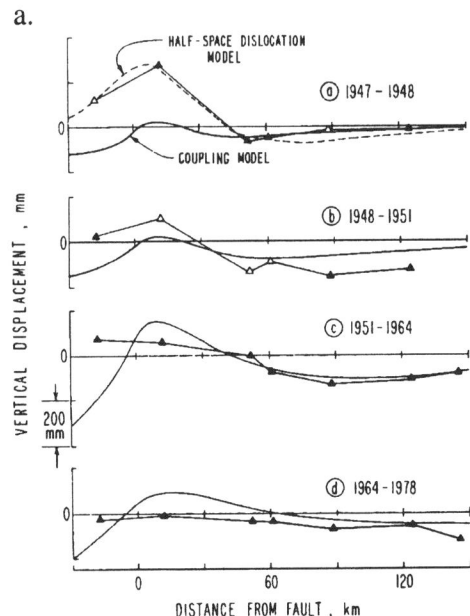

FIG. 20. (a) Vertical movement inferred from tide gauge data following 1946 Nankaido earthquake. The data is from Kato and Tsumura (1979) and Thatcher (1984); open triangles show less reliable data. The dashed line shown in the top subpanel is the prediction from a dislocation model. The solid lines are predictions from a viscoelastic earthquake model.

the Nankai Trough. There are two important features of the postseismic uplift pattern that are predicted by the model: (1) it is roughly centered over the region of coseismic subsidence, and (2) the uplift rate decreases with time following the earthquake. There are also significant differences between the predictions and the observations both in the width of the uplift feature and in the uplift amplitude at several points. Miyashita attributes the preseismic subsidence of 8.1 mm/yr at Muroto, Japan, located more than 100 km to the west of the Nankai Trough, to the 4.5-cm/yr steady-state convergence of the Philippine Sea and Eurasian plates. This steady-state convergence results in a seaward tilt of the edge of the overriding plate. Postseismic elevation changes are attributed to a combination of viscoelastic and steady-state effects, with some local uplift being due to deformation within the thin weakly coupled zone along the plate interface.

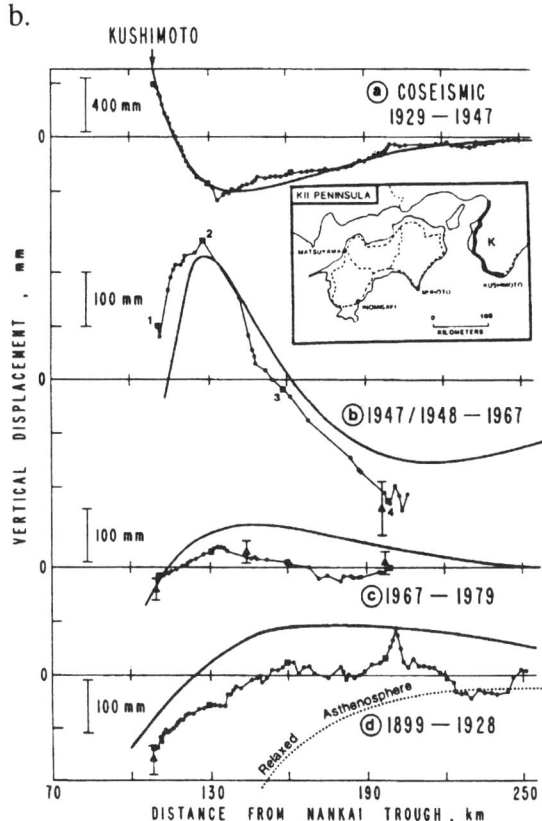

FIG. 20. (b) Vertical movement inferred from leveling surveys on the Kii Peninsula (inset) compared to viscoelastic earthquake model with coseismic slip = 4 m, dip angle = 30°, lithospheric thickness = 30 km, fault depth = 30 km, and ratio of earthquake recurrence time to Maxwell time = 40. Tide gauge observations are shown as triangles with error bars. (Both panels from Thatcher and Rundle 1984.)

Stuart (1988) found that a deformation model based on slip and velocity-dependent friction adequately described many of the temporal-spatial uplift characteristics associated with the Nankai trough deformation cycle. It does not, however, explain the long-term postseismic relaxation that Rundle, Thatcher, and Miyashita have attributed to viscoelastic effects.

Savage and Thatcher (1992) reexamined the spatial and temporal pattern of interseismic deformation at the Nankai Trough by studying tide gauge, leveling, and triangulation measurements made on Shikoku and the Kii Peninsula. The tide gauge data was well fit by a superposition of an

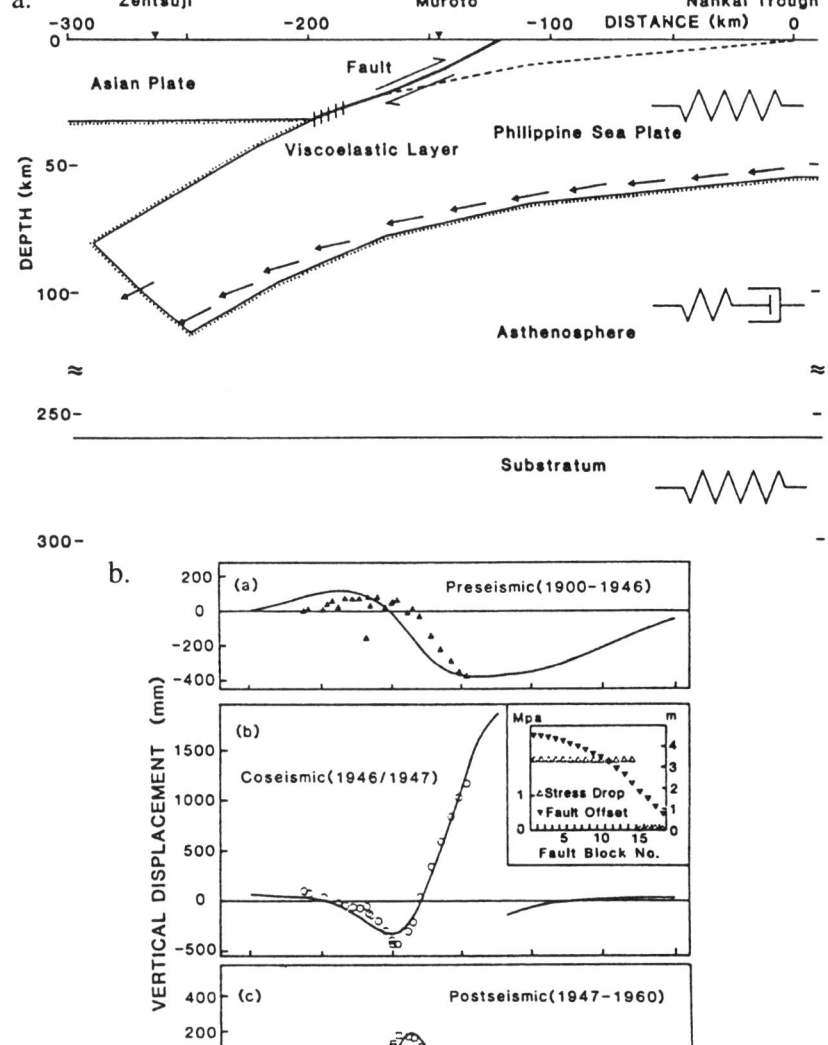

exponentially decaying transient, with a time scale of less than a decade, and a steady-state term. They examined both viscoelastic and elastic models for explaining the interseismic deformation and concluded that an elastic model was a somewhat better fit to the observations. Their elastic slip models are a bit more complex than the simple models we presented in the early sections of this article. The shallow portions of the plate interface are fully locked (with the exception of the shallowest portion in the accretionary prism) between earthquakes. The locked interface is modeled in a customary manner, that is, by imposing a virtual normal slip on the fault at the plate convergence rate. However, at intermediate depths, the plate interface is neither fully locked nor fully free to slip; hence, the interface is modeled by normal slip at a velocity less than the convergence rate. At still greater depths, the interface is free to slip, so no elastic strain accumulates due to this deep motion. One difficulty that Savage and Thatcher (1992) recognized is that the currently observed temporal and spatial pattern of deformation is different from that observed before the last earthquake. The implication is that either there must be a long-term change in the deformation pattern within an earthquake cycle or the pattern of strain accumulation and release is different from one earthquake to the next.

Savage (1995) has analyzed another five years (through 1990) of tide gauge data. He found little evidence for any temporal variation in uplift rates over the past three decades, i.e., once the early postseismic transient passed. He again argued that the observations are fit somewhat better by the elastic half-space model rather than by the viscoelastic coupling model.

Tabei *et al.* (1996) have reported on GPS measurements in Shikoku, Japan for the period from 1990 through 1995. They detected a northwest-to-southeast contraction at a rate of 2.2–3.4×10^{-7} yr^{-1} and an orientation close to that of the convergence of the Philippine plate with respect to the Eurasian plate. These results are particularly interesting when compared to the triangulation results for the same region for the period from 1949 to 1981. As deduced from the results reported by Savage and Thatcher (1992), the triangulation data gave average contraction rates over this 32-year interval that varied spatially from about $(0.9 \pm 0.9) \times 10^{-7}$ yr^{-1} to $(5.3 \pm 0.9) \times 10^{-7}$ yr^{-1}. In the triangulation data, there is a peak in the contraction rate at some distance inland from the coast. This peak is

---

FIG. 21. (a) Schematic model of 1946 Nankaido earthquake geometry. (b) Observed and predicted preseismic, coseismic, and postseismic vertical motions. (Miyashita, 1987; reproduced with permission from *J. Phys. Earth*, ©1987 Seismological Society of Japan.)

predicted by the elastic dislocation models. A similar peak is not apparent in the GPS data, although the limited number of sites in the GPS network make it difficult to draw definitive conclusions concerning spatial variations in the contraction rate.

The preceding few paragraphs have summarized some of the key studies of the earthquake cycle for the Nankai trough. It would be a prohibitively long task to review all of the numerical modeling studies of crustal deformation in Japan. However, from a historical perspective, a particularly important study of postseismic rebound is that of Thatcher *et al.* (1980). They argued that the long-term subsidence of 35 cm after 74 years over a region about 75 km wide observed following an 1896 intraplate thrust earthquake near Riku-u, Japan is consistent with viscoelastic rebound but not creep at depth. The parameters of the model were a coseismic slip of 4 m on a fault extending from the surface to 15 km depth with a dip angle of 45°, a lithospheric thickness of 30 km, and an asthenospheric viscosity of 10^{19} Pa s. This paper is one of the earliest convincing illustrations of viscoelasticity controlling postseismic crustal deformation. The parameters of the model appear to be robust, for they do not change much when the effects of relaxation in the lower crust are added to the model (Cohen, 1984).

4.2. Eastern Aleutian Subduction Zone, Southcentral Alaska

The 1964 Prince William Sound earthquake is notable both for its size (M_w = 9.2, maximum slip ~ 25 m) and for the availability of a variety of preseismic, coseismic, and postseismic deformation data. Most of the slip took place as thrust movement along the shallow previously locked Pacific–North America plate interface, but secondary, high-angle reverse slip occurred on the Patten Bay fault located inward of the Aleutian Trench. Studies by Holdahl and Sauber (1994) and Johnson *et al.* (1996) have used geodetic, geological, and tsunami observations as input to a multipatch elastic dislocation model to deduce the spatial variation in the coseismic slip. While slip occurred over a 500-km long by 200-km wide segment of the plate boundary, slip was concentrated on an easterly patch near Prince William Sound and on a secondary patch near Kodiak Island to the southwest. Brown *et al.* (1977) examined four leveling surveys conducted between 1964 and 1975 along Turnagain Arm, located to the south of Anchorage. During the first postseismic year, as much as 15 cm of relative postseismic uplift occurred along the Turnagain Arm profile. The uplift grew to between 37 and 50 cm uplift (depending on assumptions about the reference point motion) by 1975. Brown et al. (1977) proposed several possible models for the observed uplift, but preferred a deep-creep model. Comparisons of recent GPS observations with historic leveling data, as reported in Cohen et al. (1995) and Cohen and Freymueller (1997), have revealed that the postseismic uplift grew to as much as 90 cm uplift on the adjacent Kenai Peninsula sometime during the 30 years following the earthquake. Cohen (1996, 1998) studied both viscoelastic and creep models and concluded that the observed uplift is most likely due to creep. He also deduced that the rapid uplift persisted through the first postseismic year and decayed thereafter with a time scale of 3–6 yr. His preferred two-dimensional postseismic creep model, shown in Figure 22a, involves about 2.5 m slip along the down-dip extension of the coseismic rupture plane. The dip angle of not more than about 11° is shallower than that used by Brown et al. (1977), but is consistent with seismic tomography (Zhao et al. 1995). The dip angle for that portion of the plate interface that slipped postseismically is, however, greater than the dip angle for the shallower portion of the plate interface that slipped coseismically. While the two-dimensional model provides an adequate fit to the data along a profile extending inland from the plate boundary, the areal deformation pattern shown in Figure 22b reveals a complexity which is not adequately described by such a simple model.

Savage and Plafker (1991) updated the earlier study of tide gauge data for southern Alaska in the aforementioned paper by Brown et al. (1977). They found that most of the tide gauges that are located in the region of coseismic subsidence showed postseismic uplift. The uplift rates averaged over the period from 1965 to 1988 vary from less than a few millimeters per year to a few centimeters per year. To a rough approximation, the rate of uplift increased with the amount of coseismic subsidence. An exception was the site at Nikiski along Cook Inlet where the coseismic subsidence was modest but the postseismic uplift rate was substantial. Unfortunately, the tide gauge record at Nikiski is short and it is not clear whether the large uplift rate is a local anomaly or an indication of a departure from the more general trend. Savage and Plafker (1991) also found that there was a sinusoidal oscillation in the uplift rate at several locations during the first several years following the earthquake. Although they argued for a tectonic, rather than an oceanic, origin for this signal, no convincing physical explanation was advanced. While the tide gauge data averaged over the entire postseismic period through 1998 showed little evidence of the rapid decay in uplift rate that was found from the leveling data, it is curious that the uplift rates deduced by Brown et al. (1977) from tide gauge data during the first postseismic decade are much higher than those deduced over 23 years by Savage and Plafker (1991). There is also evidence for

a.

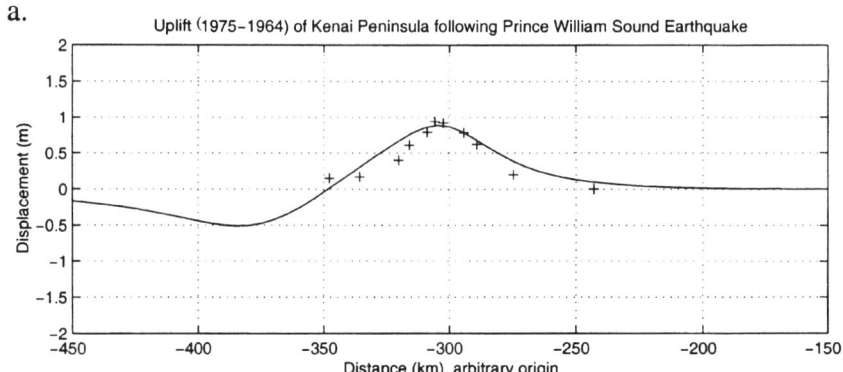

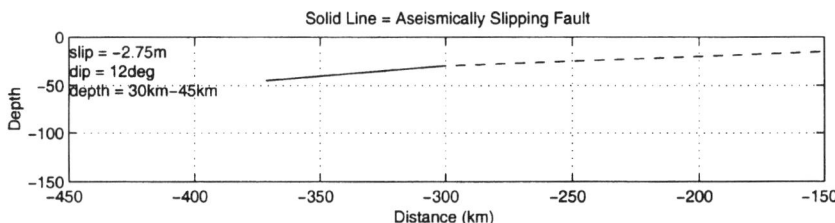

b.

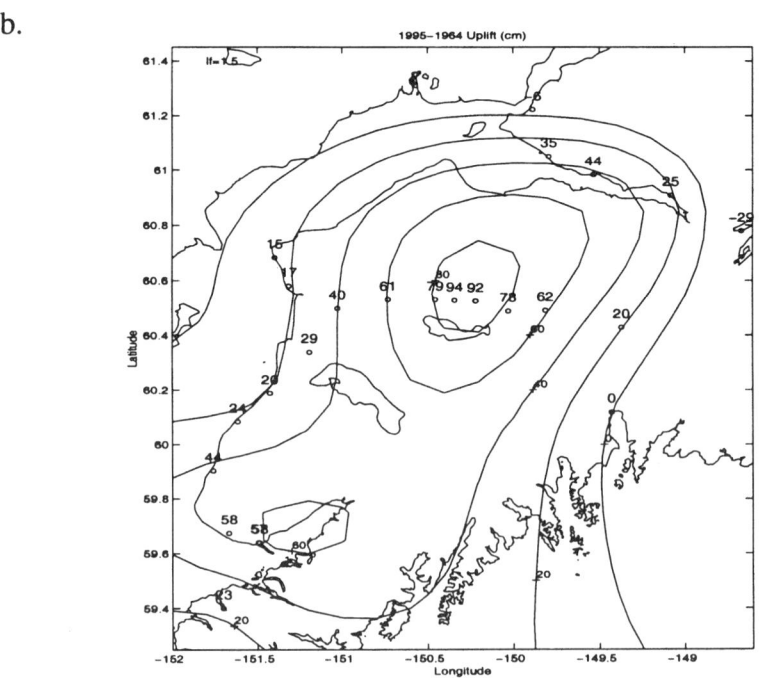

time-dependent changes in the uplift rate at Kodiak Island. Tide gauge observations made prior to the earthquake indicated uplift rates of only a few millimeters per year, while tide gauge measurements made since the earthquake (Savage and Plafker, 1991), VLBI measurements made more than two decades following the earthquake (Ryan *et al.*, 1993), and recent GPS measurements (Sauber *et al.*, 1994) all indicate uplift rates in excess of 1 cm/yr. As Savage and Plafker (1991) have commented, uplift rates of about 1 cm/yr cannot be sustained over the entire earthquake cycle. Since the geological record indicates that the recurrence interval for great earthquakes is about 750 yr while the plate convergence rate is 5.5 cm/yr, the accumulated slip deficit prior to an earthquake would be larger than the observed coseismic slip of about 20 m. The plate rate is well constrained, so either a portion of the plate convergence does not couple into megathrust earthquakes, a portion of the convergence is accommodated by motions on subsidiary structures, the recurrence interval for megathrust events is not much more than about 360 yr, or successive earthquakes have different slip amplitudes. Large earthquakes may occur between the more infrequent great earthquakes in some place in southcentral Alaska such as Kodiak Island.

Wahr and Wyss (1980) proposed a novel viscoelastic model for the observed postseismic uplift following the 1964 Prince William Sound, Alaska earthquake which provides an interesting contrast to both creep-at-depth and layered viscoelastic models. The viscoelastic flow is confined to a rectangular inclusion located just inland of the down-dip end of the coseismic rupture plane. The inclusion is 40 km wide and extends from a depth of 20 to 80 km. Confining the flow to an inclusion improves the fit between a viscoelastic model and observed postseismic uplift; otherwise, the model underpredicts the magnitude and overpredicts the width of the deformation feature. However, in order to make the model agree with both the coseismic and postseismic uplift data, the rupture plane must be extended to 400 km from the trench. This is a much wider rupture plane than is usually attributed to the Prince William Sound earthquake. Furthermore, the inclusion is not located near the volcanic regions where viscoelastic flow might otherwise be expected; therefore, the suitability of the model is questionable. Further details on comparisons of viscoelastic

---

FIG. 22. (a) Data points are a cross section of uplift data (1996–1964 height differences) for postseismic rebound on Kenai Peninsula, Alaska. The solid line is the prediction from an elastic dislocation theory calculation using the parameters shown. (b). Contours of constant uplift for the Kenai Peninsula and surrounding regions. The data used to generate the contours are also shown.

and creep-at-depth models for rebound from the Prince William Sound earthquake can be found in Cohen (1996).

4.3. San Andreas Transform Fault System, California

It would be futile to try to summarize in a few paragraphs all the work which has been done in modeling the seismotectonics in western North America or even along specific segments of the San Andreas fault system. For example, Lisowski *et al.* (1991), who developed a model for the velocity field along the San Andreas fault in central and southern California, presented a table with 17 citations to papers published between 1979 and 1991 that contain dislocation models for specific locales in California. The number of elastic dislocation and viscoelastic dislocation models has increased rapidly since then, particularly with the advent of frequent or continuous geodetic surveys using GPS. Here we comment on just a few selected illustrative results. Early attempts to model the surface displacements associated with San Andreas fault earthquakes for which there are historical observations include those for the 1934 Parkfield earthquake (Savage and Burford, 1970) and the 1906 San Francisco earthquake (Thatcher, 1974, 1975). In particular, one of the important results from the study of the San Francisco earthquake was that the slip depth was shallow, probably about 10 km in most places, although somewhat deeper slip may have been deduced if the vertical stratification in the elastic properties had been considered. Thatcher *et al.* (1997) have studied the spatial variations in the slip distribution for the San Francisco earthquake. They divided the San Andreas fault into several narrow rectangular patches, each extending to 10 km depth, and used inversion techniques with standard elastic dislocation models to determine the displacement in each patch. The result, shown in Figure 23, indicates that slip varied nonmonotonically from 8–9 m at Shelter Cove, near the northern end of the rupture, to less than 3 m at San Juan Bautista, near the southern end, with a possible intermediate peak of 8 m near Tomales Bay. An interesting feature that emerges from the analysis is that the measured surface offsets across the fault averaged about 80% of the geodetically determined slip.

Analyses by Turcotte *et al.* (1979) and Spence and Turcotte (1979) were some of the earliest attempts to model the cycle along the strike-slip San Andreas fault. Their approach to the problem was to obtain solutions to the quasi-static equilibrium equations in both the lithosphere and the asthenosphere, using Laplace transform techniques in the latter case, and to couple the equations by requiring continuity in displacement and shear stress at the lithosphere–asthenosphere boundary. (As an aside, we note

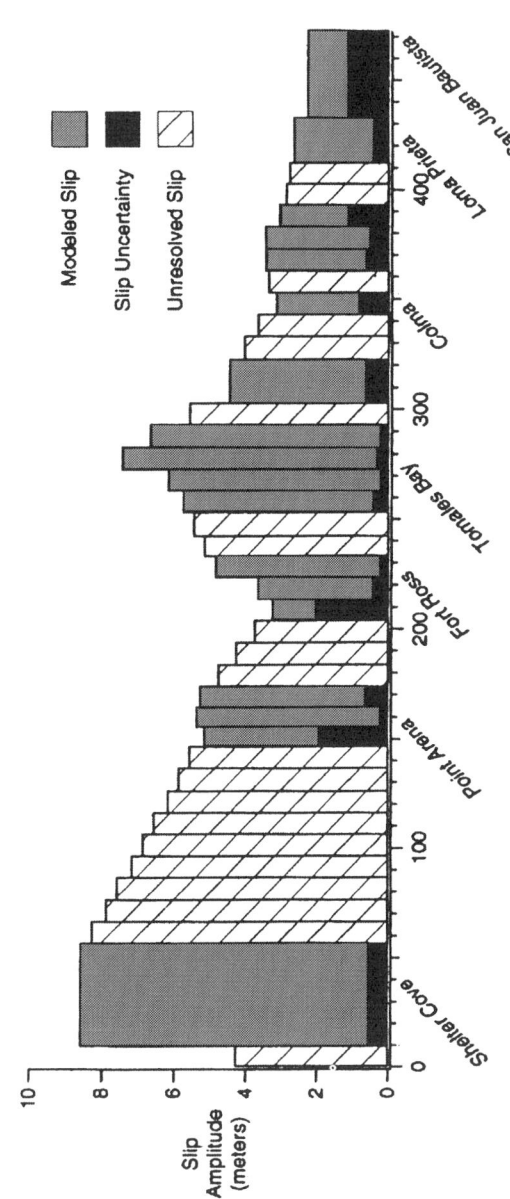

FIG. 23. Estimates of slip in 1906 San Francisco earthquake as derived from geodetic observations. (From Thatcher *et al.* 1997.)

that while it is apparent that the coupling is an essential feature of viscoelastic models, it is also important in an elastic context in overcoming defects in thin-plate models.). The deformation is driven by a periodic sequence of slip events at the fault and a uniform shear strain chosen so that the cumulative surface shear stress vanishes with the occurrence of each earthquake. The entire lithosphere is locked between earthquakes and ruptures in each event. The authors found that effects of the earthquake cycle die out with distance on a scale of a few times the lithospheric thickness. The physics involved in this model of the earthquake cycle is similar to that in Savage and Prescott (1978a) and compatible results are achieved. Rundle (1986) developed a viscoelastic model of contemporary deformation in southern California and concluded that as much as several millimeters per year of the currently observed site velocities might be due to a rebound from the last great event in southern California in 1857. This conclusion was based on an assumed viscosity between 5×10^{18} and 1×10^{20} Pa s and an elastic thickness of 20–40 km, with the lower range for the thickness being preferred. Kroger *et al.* (1987) used viscoelastic models to examine Very Long Baseline Interferometry observations in California. The model had a viscoelastic asthenosphere, and the model represented crustal faults as thin zones of viscoelastic weakness with depth-dependent viscosities. Li and Lim (1988) extended the aforementioned work of Li and Rice (1987) to model VLBI and triangulation observations in southern California, specifically in the area of the San Andreas and San Jacinto faults. The parameters included the depth of fault locking, 8–11 km; elastic lithosphere thicknesses, 20–22.5 km; earthquake recurrence time, 250 yr on the San Andreas fault and 100 yr on the San Jacinto fault; lower crust or upper mantle viscosity, $2-10 \times 10^{18}$ Pa s, and loading rates, 3.0–3.5 cm/yr on the San Andreas fault and 1.0–1.3 cm/yr on the San Jacinto fault. The authors commented that while they were able to fit the available geodetic data, the model parameters were not tightly resolved.

Postseismic deformation has been observed after many of the major shocks occurring recently in California, including the 1983 Coalinga ($M_w = 6.9$), the 1989 Loma Prieta ($M_w = 7.0$), the 1992 Landers ($M_w = 7.3$), and the 1994 ($M_w = 6.7$) Northridge earthquakes. Postseismic deformations lasting for time scales ranging from a few weeks to several years have been observed. Massonnet *et al.* (1996) and Savage and Svarc (1997) have used aseismic slip models to explain the multiyear postseismic rebound for the Landers event, although the slip depths of 6–11 km used by the former authors are shallower than 10–30 km preferred by the latter.

Peltzer *et al.* (1977) disagreed with the model of Savage and Svarc (1997) and attributed more of the deformation signal to shallow slip. They also attributed short-term (~ 8 months) relaxation near step-overs in the coseismic rupture pattern to fluid flow due to pore pressure changes generated by the earthquake. Shen *et al.* (1994) modeled the short-term (several weeks) relaxation following the Landers event by a combination of slip on shallow and deep dislocation surfaces. Burgmann *et al.* (1997) attributed the Loma Prieta postseismic deformation to oblique reverse slip along either the Loma Prieta rupture or the San Andreas and thrust faulting along the nearby Foothills fault system. Hager *et al.* (1997) argued that the relatively deep coseismic slip for the Northridge earthquake, extending to 20 km depth, suggests that an elastic over viscoelastic model is more appropriate than an elastic half-space model. Within the context of their model, the high strain rates that were observed within the Ventura basin require that depth variations in the elastic constants be taken into account. Shen *et al.* (1996) observed that the fault parallel displacement field in southern California is broader than that predicted by elastic dislocation models and speculated that the possible causes for the broadening are motion along a detachment fault in the lower crust or upper mantle, viscoelastic flow below the seismogenic layer, or some unspecified broad deformation across the San Andreas fault.

5. Epilogue

In this article, I have tried to find a middle ground between reviewing only a few topics in great detail or many topics superficially. It might be helpful, at the conclusion, to summarize what I believe are the most salient general topics that have been reviewed. From the standpoint of the physics of the crustal deformation process, the most important topics are the representation of fault slip as a dislocation, the modeling of crustal deformation as a quasi-equilibrium process, the development of a kinematically self-consistent model of the earthquake cycle, and the evaluation of the role played by rheology and creep processes in transferring stress both vertically and horizontally. The key mathematical tools that come into play are techniques for solving differential equations including Green's functions and direct approaches, matrix propagators, Laplace and Fourier transforms, and finite-element methods. Using these ideas and techniques, scientists have progressed in their understanding of crustal deformation processes from simple models of the coseismic slip in a large earthquake to detailed models of the entire earthquake cycle. In favorable cases, it is now possible to study the spatial details of fault-slip patterns, to determine subsurface fault creep rates, and to estimate the viscosity of the lower crust or upper mantle. Nevertheless, many fundamental and practical issues remain to be studied. For example, it remains difficult to unambiguously distinguish between deep fault creep and viscoelastic flow, to model the evolution of the deformation signal from the earthquake cycle to geological time scales, or even to fully resolve all parameters for a particular physical model. More work is required to integrate fault mechanics, particularly various slip criteria, into the numerical models. While geodetic data has been and will likely continue to be the most important source of information to be used in constraining crustal deformation models, increasingly sophisticated seismological, geological, rock mechanical, and other geophysical constraints will also be required.

Acknowledgments

The author wishes to thank Desmond Darby, Jeff Freymueller, Jeanne Sauber, Mark Taylor, Jim Savage, and an anonymous reviewer for their insightful scientific comments on preliminary drafts of this paper. He also expresses appreciation to Davria Cohen for her help in editing the initial manuscript.

Appendix

Elastic Displacement Due to a Point Source

Using the coordinate system shown in Figure 24, the equations published by Okada (1992) for the elastic displacement due to a point source with seismic moment, M_0, and dip angle, δ, located at $(0, 0, -c)$ can be written in the form

$$u(x, y, z) = \frac{M_0}{2\pi\mu} \left[u_A^0(x, y, z) - u_A^0(x, y, -z) + u_B^0(x, y, z) + z u_C^0(x, y, z) \right],$$

where u_A^0, u_B^0, and u_C^0 are given in Table A1.

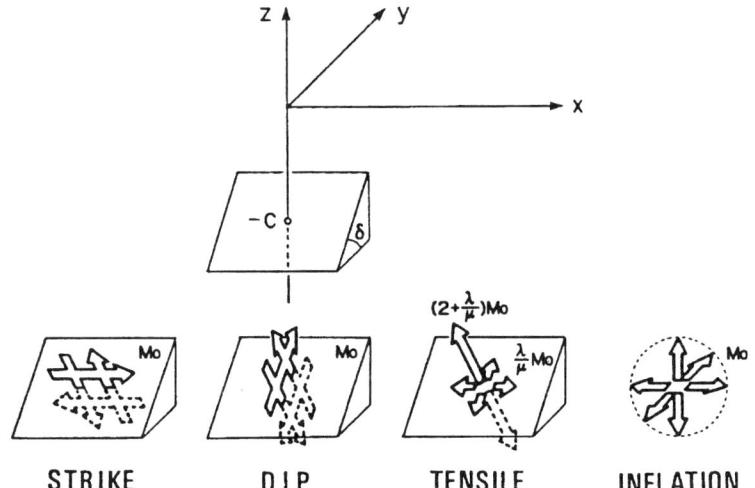

FIG. 24. Coordinate system used for the elastic dislocation equations of Okada [1985, 1992]. (Okada, 1992; reproduced with permission from *Bull. Seismol. Soc. Am.*, ©1992 Seismological Society of America.)

TABLE A1

Type	u_A^0	u_B^0	u_C^0
Strike	$\dfrac{1-\alpha}{2}\dfrac{q}{R^3}+\dfrac{\alpha}{2}\left(\dfrac{3x^2q}{R^5}\right)$ $\dfrac{1-\alpha}{2}\dfrac{x}{R^3}\sin\delta+\dfrac{\alpha}{2}\left(\dfrac{3xyq}{R^5}\right)$ $-\left(\dfrac{1-\alpha}{2}\right)\dfrac{x}{R^3}\cos\delta+\dfrac{\alpha}{2}\left(\dfrac{3xdq}{R^5}\right)$	$-\dfrac{3x^2q}{R^5}-\dfrac{1-\alpha}{\alpha}I_1^0\sin\delta$ $-\dfrac{3xyq}{R^5}-\dfrac{1-\alpha}{\alpha}I_2^0\sin\delta$ $-\dfrac{3cxq}{R^5}-\dfrac{1-\alpha}{\alpha}I_4^0\sin\delta$	$-(1-\alpha)\dfrac{A_3}{R^3}\cos\delta+\alpha\dfrac{3cq}{R^5}A_5$ $-(1-\alpha)\dfrac{3xy}{R^5}\cos\delta+\alpha\dfrac{3cx}{R^5}\left[\sin\delta-\dfrac{5yq}{R^2}\right]$ $-(1-\alpha)\dfrac{3xy}{R^5}\sin\delta+\alpha\dfrac{3cx}{R^5}\left[\cos\delta+\dfrac{5dq}{R^2}\right]$
Dip-slip	$\dfrac{\alpha}{2}\left(\dfrac{3xpq}{R^5}\right)$ $\dfrac{1-\alpha}{2}\dfrac{s}{R^3}+\dfrac{\alpha}{2}\left(\dfrac{3ypq}{R^5}\right)$ $-\dfrac{1-\alpha}{2}\dfrac{t}{R^3}+\dfrac{\alpha}{2}\left(\dfrac{3dpq}{R^5}\right)$	$-\dfrac{3xpq}{R^5}+\dfrac{1-\alpha}{\alpha}I_3^0\sin\delta\cos\delta$ $-\dfrac{3ypq}{R^5}+\dfrac{1-\alpha}{\alpha}I_1^0\sin\delta\cos\delta$ $-\dfrac{3cpq}{R^5}+\dfrac{1-\alpha}{\alpha}I_5^0\sin\delta\cos\delta$	$(1-\alpha)\dfrac{3xt}{R^5}-\alpha\dfrac{15cxpq}{R^7}$ $-(1-\alpha)\dfrac{1}{R^3}\left[\cos2\delta-\dfrac{3yt}{R^2}\right]+\alpha\dfrac{3c}{R^5}\left[s-\dfrac{5ypq}{R^2}\right]$ $-(1-\alpha)\dfrac{A_3}{R^3}\sin\delta\cos\delta-\alpha\dfrac{3c}{R^5}\left[t+\dfrac{5dpq}{R^2}\right]$
Tensile	$\dfrac{1-\alpha}{2}\dfrac{x}{R^3}-\dfrac{\alpha}{2}\left(\dfrac{3xq^2}{R^5}\right)$ $\dfrac{1-\alpha}{2}\dfrac{t}{R^3}-\dfrac{\alpha}{2}\left(\dfrac{3yq^2}{R^5}\right)$ $\dfrac{1-\alpha}{2}\dfrac{s}{R^3}-\dfrac{\alpha}{2}\left(\dfrac{3dq^2}{R^5}\right)$	$\dfrac{3xq^2}{R^5}-\dfrac{1-\alpha}{\alpha}I_3^0\sin^2\delta$ $\dfrac{3yq^2}{R^5}-\dfrac{1-\alpha}{\alpha}I_1^0\sin^2\delta$ $\dfrac{3cq^2}{R^5}-\dfrac{1-\alpha}{\alpha}I_5^0\sin^2\delta$	$-(1-\alpha)\dfrac{3xs}{R^5}+\alpha\dfrac{15cxq^2}{R^7}-\alpha\dfrac{3xz}{R^5}$ $(1-\alpha)\dfrac{1}{R^3}\left[\sin2\delta-\dfrac{3ys}{R^2}\right]+\alpha\dfrac{3c}{R^5}\left[t-y+\dfrac{5yq^2}{R^2}\right]-\alpha\dfrac{3yz}{R^5}$ $-(1-\alpha)\dfrac{1}{R^3}[1-A_3\sin^2\delta]-\alpha\dfrac{3c}{R^5}\left[s-d+\dfrac{5dq^2}{R^2}\right]+\alpha\dfrac{3dz}{R^5}$
Inflation	$-\dfrac{1-\alpha}{2}\dfrac{x}{R^3}$ $-\dfrac{1-\alpha}{2}\dfrac{y}{R^3}$ $-\dfrac{1-\alpha}{2}\dfrac{d}{R^3}$	$\dfrac{1-\alpha}{\alpha}\dfrac{x}{R^3}$ $\dfrac{1-\alpha}{\alpha}\dfrac{y}{R^3}$ $\dfrac{1-\alpha}{\alpha}\dfrac{d}{R^3}$	$(1-\alpha)\dfrac{3xd}{R^5}$ $(1-\alpha)\dfrac{3yd}{R^5}$ $(1-\alpha)\dfrac{C_3}{R^3}$

Some supplemental definitions are:

$d = c - z$

$R^2 = x^2 + y^2 + d^2$

$\alpha = \dfrac{\lambda + \mu}{\lambda + 2\mu}$

$I_1^0 = y\left[\dfrac{1}{R(R+d)^2} - x^2 \dfrac{(3R+d)}{R^3(R+d)^3}\right]$

$I_2^0 = x\left[\dfrac{1}{R(R+d)^2} - y^2 \dfrac{(3R+d)}{R^3(R+d)^3}\right]$

$I_3^0 = \dfrac{x}{R^3} - I_2^0$

$I_4^0 = -xy\dfrac{2R+d}{R^3(R+d)^2}$

$I_5^0 = \dfrac{1}{R(R+d)} - x^2\dfrac{2R+d}{R^3(R+d)^2}$

$p = y\cos\delta + d\sin\delta$

$q = y\sin\delta - d\cos\delta$

$s = p\sin\delta + q\cos\delta$

$t = p\cos\delta - q\sin\delta$

$A_3 = 1 - \dfrac{3x^2}{R^2}$

$A_5 = 1 - \dfrac{5x^2}{R^2}$

$C_3 = 1 - \dfrac{3d^2}{R^2}$

References

Abe, K. (1977). Tectonic implications of the large Skioya-Oki earthquakes of 138. *Tectonophysics* **41**, 269–289.

Abramowitz, M., and Stegun, I. A. (1972). "Handbook of Mathematical Functions." Dover, New York.

Aki, K., and Richards, P. G. (1980). "Quantitative Seismology Theory and Methods." W. H. Freeman and Co., San Francisco.

Akin, J. E. (1982). "Application and Implementation of Finite Element Methods." Academic Press Limited, London.

Ando, M. (1975). Source mechanisms and tectonic significance of historical earthquakes along the Nankai rough, Japan. *Tectonophysics* **27**, 119–140.

Barnett, D. M., and Freund, L. B. (1975). An estimate of strike-slip fault friction stress and fault depth from surface displacement data. *Bull. Seismol. Soc. Am.* **65**, 1254–1266.

Bathe, K.-J. (1982). "Finite-Element Procedures in Engineering Analysis." Prentice-Hall, Englewood Cliffs, NJ.

Beer, G., and Watson, J. O. (1994). "Introduction to Finite and Boundary Element Methods for Engineers." John Wiley & Sons, Chichester.

Ben-Menahem, A., and Singh, S. J. (1968). Multipolar elastic fields in a layered half-space. *Bull Seismol. Soc. Am.* **58**, 1519–1572.

Ben-Menahem, A., and Singh, S. J. (1970). Deformation of a spherical Earth model by finite dislocations. In "Earthquake Displacement Fields and the Rotation of the Earth" (L. Mansinha *et al.*, eds.), Reidel, Dordrecht, pp. 39–42.

Ben-Menahem, A., Singh, S. J., and Solomon, F. (1970). Deformation of a homogeneous Earth model by finite dislocations. *Rev. Geophys. Space Phys.* **8**, 591–632.

Bischke, R. E. (1976). A model of convergent plate margins based on the recent tectonics of Shikoku, Japan. *J. Geophys. Res.* **79**, 4845–4857.

Bonafede, M. (1990). Axi-symmetric deformation of a thermo-poro-elastic half-space: inflation of a magma chamber. *Geophys. J. Int.* **103**, 289–299.

Bonafede, M., Dragoni, M., and Quareni, F. (1986). Displacement and stress fields produced by a centre of dilation and by a pressure source in a viscoelastic half-space. *Geophys. J. Roy. Astron. Soc.* **87**, 455–485.

Brown, L. D., Reilinger, R. E., Holdahl, S. R., and Balazs, E. I. (1977). Postseismic crustal uplift near Anchorage, Alaska. *J. Geophys. Res.* **82**, 3369–3378.

Burgmann, R., Segall, P., Lisowski, M., and Svarc, J. (1997). Postseismic strain following the 1989 Loma Prieta earthquake from GPS and leveling measurements. *J. Geophys. Res.* **102**, 4933–4955.

Chinnery, M. A. (1961). The deformation of the ground around surface faults. *Bull. Seismol. Soc. Am.* **50**, 355–372.

Chinnery, M. A. (1963). The stress changes that accompany strike-slip faulting. *Bull. Seismol. Soc. Am.* **53**, 921–932.

Chinnery, M. A., and Jovanovich, D. B. (1972). Effect of Earth layering on earthquake displacement fields. *Bull. Seismol. Soc. Am.* **62**, 1629–1639.

Cohen, S. C. (1982). A multilayer model of time dependent deformation following an earthquake on a strike-slip fault. *J. Geophys. Rev.* **87**, 5409–5421.

Cohen, S. C. (1984). Postseismic deformation due to subcrustal viscoelastic relaxation following dip-slip earthquakes. *J. Geophys. Res.* **89**, 4538–4544.

Cohen, S. C. (1992). Postseismic deformation and stress diffusion due to viscoelasticity and comments on the modified Elsasser model. *J. Geophys. Res.* **97**, 15395–15403.

Cohen, S. C. (1994). Evaluation of the importance of model features for cyclic deformation due to dip-slip faulting. *Geophys. J. Int.* **119**, 831–841.

Cohen, S. C. (1996). Time-dependent uplift of the Kenai Peninsula and adjacent regions of south central Alaska since the 1964 Prince William Sound Earthquake. *J. Geophys. Res.* **101**, 8395–8604.

Cohen, S. C. (1998). On the rapid postseismic uplift along Turnagain Arm, Alaska following the 1964 Prince William Sound Earthquake. *Geophys. Res. Lett.* **25**, 1213–1215.

Cohen, S. C., and Freymueller, J. T. (1997). Deformation of the Kenai Peninsula, Alaska. *J. Geophys. Res.* **102**, 20479–20487.

Cohen, S. C., and Kramer, M. J. (1984). Crustal deformation, the earthquake cycle, and models of viscoelastic flow in the asthenosphere. *Geophys. J. Roy. Astron. Soc.* **78**, 735–750.

Cohen, S., Holdahl, S., Caprette, D., Hilla, S., Safford, R., and Schultz, D. (1995). Uplift of the Kenai Peninsula, Alaska, since the 1964 Prince William Sound Earthquake. *J. Geophys. Res.* **100**, 2031–2038.

Converse, G., and Comninou, M. (1975). Dependence on the elastic constants of surface deformation due to faulting. *Bull. Seismo. Soc. Am.* **65**, 1173–1176.

Dmowska, R., Rice, J. R., Lovison, L. C., and Josell, D. (1988). Stress transfer and seismic phenomena in coupled subduction zones during the earthquake cycle. *J. Geophys. Res.* **93**, 7869–7884.

Dmowska, R., Zheng, G., and Rice, J. R. (1996). Seismicity and deformation at convergent margins due to heterogeneous coupling. *J. Geophys. Res.* **101**, 3015–3029.

Douglass, J. J., and Buffett, B. A. (1995). The stress state implied by dislocation models of subduction deformation. *Geophys. Res. Lett.* **22**, 3115–3118.

Dragoni, M., Yuen, D. A., and Boschi, E. (1983). Global postseismic deformation in a stratified viscoelastic Earth. *J. Geophys. Res.* **88**, 2240–2250.

Dziewonski, A. M., and Anderson, D. L. (1981). Preliminary reference earth model. *Phys. Earth Planet. Inter.* **25**, 297–356.

Elsasser, W. M. (1969). Convection and stress propagation in the upper mantle. In "The Applications of Modern Physics to the Earth and Planetary Interiors" (S. K. Runcorn, ed.), J. Wiley and Sons, New York, pp. 223–246.

Eshelby, J. D. (1973). Dislocation theory for geophysical applications. *Philos. Trans. Roy. Soc. London Ser. A* **274**, 331–338.

Fernandez, J., Yu, T.-T., and Rundle, J. B. (1996). Horizontal viscoelastic-gravitational displacement due to a rectangular dipping thrust fault in a layered Earth model. *J. Geophys. Res.* **101**, 13581–13594.

Fitch, T. J., and Scholz, C. H. (1971). Mechanism of underthrusting in southwest Japan: a model of convergent plate interactions. *J. Geophys. Res.* **76**, 7260–7292.

Fluck, P., Hyndman, R. D., and Wang, K. (1997). Three-dimensional dislocation model for great earthquakes of the Cascadia subduction zone. *Geophys. Res.* **102**, 20539–20550.

Freund, L. B., and Barnett, D. M. (1976). A two-dimensional analysis of surface deformation due to dip-slip faulting. *Bull. Seismol. Soc. Am.* **66**, 667–675.

Gilbert, F., and Dziewonski, A. (1975). An application of normal mode theory to the retrieval of structural parameters and source mechanisms from seismic spectra. *Philos. Trans. Roy. Soc. London Ser. A* **278**, 187–269.

Hager, B., Lyzenga, G. L., Donnellan, A., and Dong, D. (1977). Reconciling rapid strain accumulation with deep earthquakes in the Ventura Basin, California (abstract). *EOS, Trans. AGU* **78**, F155.

Hodgkinson, K. M., Stein, R. S., and King, G. C. P. (1996). The 1954 Rainbow Mountain–Fairview Peak–Dixie Valley earthquakes: a triggered normal faulting sequence. *J. Geophys. Res.* **101**, 25459–25471.

Hofton, M. A., Rundle, J. B., and Foulger, G. R. (1995). Horizontal surface deformation due to dike emplacement in an elastic-gravitational layer overlying a viscoelastic-gravitational half-space. *J. Geophys. Res.* **100**, 6329–6338.

Holdahl, S. R., and Sauber, J. (1994). Coseismic slip in the 1964 Prince William Sound earthquake: a new geodetic inversion. *Pure Appl. Geophys.* **142**, 55–82.

Iwasaki, T., and Matsu'ura, M. (1981). Quasi-static strains and tilts due to faulting in a layered half-space with an intervenient viscoelastic layer. *J. Phys. Earth* **29**, 499–518.

Iwasaki, T., and Sato, R. (1979). Strain field in a semi-infinite medium due to an inclined rectangular fault. *J. Phys. Earth* **27**, 285–314.

Jaeger, J. C., and Cook, W. G. W. (1976). "Fundamentals of Rock Mechanics." Chapman and Hall, London, p. 211.

Johnson, J. M., Satake, K., Holdahl, S. R., and Sauber, J. (1996). The 1964 Prince William Sound earthquake: joint inversion of tsunami and geodetic data. *J. Geophys. Res.* **101**, 523–532.

Jovanovich, D. B. (1975). An inversion method for estimating the source parameters of seismic and aseismic events from static strain data. *Geophys. J. Roy. Astron. Soc.* **43**, 347–365.

Jovanovich, D. B., Husseini, M. I., and Chinnery, M. A. (1974a). Elastic dislocations in a layered half-space—I. Basic theory and numerical methods. *Geophys. J. Roy. Astron. Soc.* **39**, 205–217.

Jovanovich, D. B., Husseini, M. I., and Chinnery, M. A. (1974b). Elastic dislocations in a layered half-space—II. The point source. *Geophys. J. Roy. Astron. Soc.* **39**, 219–239.

Kasahara, K. (1957). The nature of seismic origins as inferred from seismological and geodetic observations (1). *Bull. Earthquake Res. Inst. Tokyo Univ.* **35**, 473–532.

Kasahara, K. (1981). "Earthquake Mechanics." Cambridge Univ. Press, Cambridge.

Kato, T., and Tsumura, K. (1979). Vertical land movement in Japan as deduced from tidal record. *Bull. Earthquake Res. Inst. Tokyo Univ.* **54**, 559–628.

King, G. C. P., Stein, R. S., and Rundle, J. B. (1988). The growth of geological structure by repeated earthquakes 1. Conceptual framework. *J. Geophys. Res.* **93**, 13307–13318.

Kittel, C. (1971). "Introduction to Solid State Physics," 4th ed. John Wiley and Sons, Inc., New York.

King, G. C. P., Stein, R. S., and Lin, J. (1997). Static stress changes and the triggering of earthquakes, preprint.

Koseluk, R. A., and Bischke, R. E. (1981). An elastic rebound model for normal fault earthquakes. *J. Geophys. Res.* **86**, 1081–1090.

Kozuch, M. J., Yu, Y.-T., and Rundle, J. B. (1996). Southeastern Caribbean sea level variability and viscoelastic relaxation. *J. Geophys. Res.* **101**, 8579–8593.

Kroger, P. M., Lyzenga, G. A., Wallace, K. S., and Davidson, J. M. (1987). Tectonic motion in the western United States from Very Long Baseline Interferometry measurements. *J. Geophys. Res.* **92**, 14151–14163.

Lay, T., and Wallace, T. C. (1995). "Modern Global Seismology." Academic Press, San Diego, pp. 319–331.

Lehner, F. K., and Li, V. C. (1982). Large-scale characteristics of plate boundary deformations relative to the post-seismic readjustment of a thin asthenosphere. *Geophys. J. Roy. Astron. Soc.* **71**, 775–792.

Lehner, F. K., Li, V. C., and Rice, J. R. (1981). Stress diffusion along rupturing plate boundaries. *J. Geophys. Res.* **86**, 6155–6169.

Li, V. C., and Lim, H. S. (1988). Surface deformations at complex strike-slip boundaries. *J. Geophys. Res.* **93**, 7943–7954.

Li, V. C., and Rice, J. R. (1987). Crustal deformation in great California earthquake cycles. *J. Geophys. Res.* **92**, 11533–11551.

Lisowski, M., Savage, J. C., and Prescott, W. H. (1991). The velocity field along the San Andreas fault in central and southern California. *J. Geophys. Res.* **96**, 8369–8389.

Lliboutry, L. A. (1987). "Very Slow Flows of Solids Basics of Modeling in Geodynamics and Glaciology." Nijhoff, Dordrecht.

Lorenzetti, E., and Tullis, T. E. (1989). Geodetic predictions of a strike-slip fault model: implications for intermediate- and short-term earthquake prediction. *J. Geophys. Res.* **94**, 12341–12361.

Lyzenga, G. A., Raefsky, A., and Mulligan, S. G. (1991). Models of recurrent strike-slip earthquake cycles and the state of crustal stress. *J. Geophys. Res.* **96**, 21623–21640.

Ma, X. Q., and Kusznir, N. J. (1992). 3-D subsurface displacement and strain fields for faults and fault arrays in a layered elastic half-space. *Geophys. J. Int.* **111**, 542–558.

Ma, X. Q., and Kusznir, N. J. (1994). Effects of rigidity layering, gravity, and stress relaxation on 3-D subsurface fault displacement field. *Geophys. J. Int.* **118**, 201–229.

Mahrer, K. D. (1984). Approximating surface deformation from a buried strike-slip fault or shear crack in a mildly uneven half-space. *Bull. Seismol. Soc. Am.* **74**, 797–803.

Mahrer, K. D., and Nur, A. (1979a). Strike slip faulting in a downward varying crust. *J. Geophys. Res.* **84**, 2296–2302.

Mahrer, K. D., and Nur, A. (1979b). Static strike-slip faulting in a horizontally varying crust. *Bull. Seismol. Soc. Am.* **69**, 975–1009.

Mansinha, L., and Smylie, D. E. (1967). Effect of earthquakes on the Chandler wobble and the secular polar shift. *J. Geophys. Res.* **72**, 4731–4743.

Mansinha, L., and Smylie, D. E. (1971). The displacement field of inclined faults. *Bull. Seismol. Soc. Am.* **61**, 1433–1449.

Mansinha, L., Smylie, D. E., and Chapman, C. H. (1979). Seismic excitation of the Chandler wobble revisited. *Geophys. J. Roy. Astron. Soc.* **59**, 1–17.

Maruyama, T. (1964). Statical elastic dislocation in an infinite and semi-infinite medium. *Bull. Earthquake Res. Inst. Tokyo University* **42**, 289–368.

Maruyama, T. (1973). Theoretical model of seismic faults. In "Publications for the 50th Anniversary of the Great Kanto Earthquake, 1923." Earthquake Research Institute, Tokyo University, Tokyo.

Massonnet, D., Thatcher, W., and Vadon, H. (1996). Radar interferometry detects two mechanism of postseismic relaxation following the Landers, California earthquake. *Nature* **382**, 612–616.

Matsu'ura, M., and Sato, R. (1978). Static deformation due to the fault spreading over several layers in a multi-layered medium. Part II: Strain and tilt. *J. Phys. Earth* **23**, 1–29.

Matsu'ura, M., and Sato, R. (1989). A dislocation model for the earthquake cycle at convergent plate boundaries. *Geophys. J. Int.* **96**, 23–32.

Matsu'ura, M., and Tanimoto, T. (1980). Quasi-static deformations due to an inclined, rectangular fault in a viscoelastic half-space. *J. Phys. Earth* **28**, 103–118.

Matsu'ura, M., Tanimoto, T., and Iwaskai, T. (1981). Quasi-static displacement due to faulting in a layered half-space with an intervenient viscoelastic layer. *J. Phys. Earth* **29**, 23–54.

Matthews, M. V., and Segall, P. (1993). Estimation of depth-dependent fault slip from measured surface deformation with application to the 1906 San Francisco earthquake. *J. Geophys. Res.* **98**, 12153–12163.

McHugh, S., and Johnston, M. (1977). Surface shear stress, strain, and shear displacement for screw dislocations in a vertical slab with shear modulus contrast. *Geophys. J. Roy. Astron. Soc.* **49,** 713–722.

Melosh, H. J. (1976). Nonlinear stress propagation in the Earth's upper mantle. *J. Geophys. Res.* **81,** 5621–5632.

Melosh, H. J. (1978). Reply to Comment on 'Nonlinear stress propagation in the earth's upper mantle by H. J. Melosh' (Savage and Prescott, 1978b). *J. Geophys. Res.* **83,** 5009–5010.

Melosh, H. J. (1983). Vertical movements following a dip-slip earthquake. *Geophys. Res. Lett.* **10,** 47–50.

Melosh, H. J., and Fleitout, L. (1982). The earthquake cycle in subduction zones. *Geophys. Res. Lett.* **9,** 21–24.

Melosh, H. J., and Raefsky, A. (1980). The dynamical origin of subduction zone topography. *Geophys. J. Roy. Astron. Soc.* **60,** 333–354.

Melosh, H. J., and Raefsky, A. (1981). A simple and efficient method for introducing faults into finite element computations. *Bull. Seismol. Soc. Am.* **71,** 1391–1400.

Melosh, H. J., and Raefsky, A. (1983). Anelastic response of the earth to a dip-slip earthquake. *J. Geophys. Res.* **88,** 515–526.

Melosh, H. J., and Williams, Jr., C. A. (1989). Mechanics of graben formation in crustal rocks: a finite element analysis. *J. Geophys. Res.* **94,** 13961–13973.

Mindlin, R. D., and Cheng, D. H. (1950). Nuclei of strain in the semi-infinite solid. *J. Appl. Phys.* **21,** 926–933.

Miyashita, K. (1987). A model of plate convergence in southwest Japan, inferred from leveling data associated with the 1946 Nankaido earthquake. *J. Phys. Earth* **35,** 449–467.

Nur, A., and Booker, J. R. (1972). Aftershocks caused by pore fluid flow? *Science* **175,** 885–887.

Nur, A. and Mavko, G. (1974). Postseismic viscoelastic rebound. *Science* **183,** 204–206.

Okada, Y. (1985). Surface deformation due to shear and tensile faults in a half-space. *Bull. Seismol. Soc. Am.* **75,** 1135–1154.

Okada, Y. (1992). Internal deformation due to shear and tensile faults in a half-space. *Bull Seismol. Soc. Am.* **82,** 1018–1040.

Okubo, S. (1993). Reciprocity theorem to compute the static deformation due to a point dislocation buried in a spherically symmetric earth. *Geophys. J. Int.* **115,** 921–928.

Pacheco, J. F., Sykes, L. R., and Scholz, C. H. (1993). Nature of seismic coupling along simple plate boundaries of the subduction type. *J. Geophys. Res.* **98,** 14133–14159.

Peltzer, G., Rosen, P., Rogez, F., and Hudnut, K. (1977). Post-seismic transients after the 1992 Landers earthquake observed with SAR interferometry (abstract). *EOS, Trans. AGU* **78,** F165.

Piersanti, A., Spada, G., and Sabadini, R. (1997). Global postseismic rebound of a viscoelastic earth: theory for finite faults and application to the 1964 Alaska earthquake. *J. Geophys. Res.* **102,** 477–492.

Piersanti, A., Spada, G., Sabadini, R., and Bonafede, M. (1995). Global post-seismic deformation. *Geophys. J. Int.* **120,** 544–566.

Plafker, G. (1965). Tectonic deformation associated with the 1964 Alaska earthquake. *Science* **148,** 1675–1687.

Pollitz, F. F. (1992). Postseismic relaxation theory on the spherical earth. *Bull. Seismol. Soc. Am.* **82,** 422–453.

Pollitz, F. F. (1996). Coseismic deformation from earthquake faulting on a layered spherical earth. *Geophys. J. Int.* **125,** 1–14.

Pollitz, F. F. (1997). Gravitational viscoelastic postseismic relaxation on a layered spherical earth. *J. Geophys. Res.* **102,** 17921–17941.

Press, F. (1965). Displacements, strains, and tilts at teleseismic distance. *J. Geophys. Res.* **70,** 2395–2412.

Rani, S., and Singh, S. J. (1992). Static deformation of a uniform half-space due to a long dip-slip fault. *Geophys. J. Int.* **109,** 469–476.

Reches, Z., Schubert, G., and Anderson, C. (1994). Modeling of periodic earthquakes on the San Andreas fault: effects of nonlinear crustal rheology. *J. Geophys. Res.* **99,** 21983–22000.

Reid, H. F. (1911). The elastic rebound theory of earthquakes. *Bull. Dept. Geol. Univ. Calif.* **6,** 413–444.

Rice, J. R. (1968). In "Fracture: An Advanced Treatise," Vol. II (H. Liebowitz, ed.), Academic Press, New York.

Rongved, L., and Frasier, J. T. (1958). Displacement discontinuity in the elastic half-space. *J. Appl. Mech.* **25,** 125–128.

Rosenman, M., and Singh, S. J. (1973a). Quasi-static strains and tilts due to faulting a viscoelastic half-space. *Bull. Seismol. Soc. Am.* **63,** 1737–1752.

Rosenman, M., and Singh, S. J. (1973b). Stress relaxation in a semi-infinite viscoelastic earth model. *Bull. Seismol. Soc. Am.* **63,** 2145–2154.

Roth, F. (1990). Subsurface deformations in a layered elastic half-space. *Geophys. J. Int.* **103,** 147–155.

Rundle, J. B. (1976). Anelastic processes in strike slip faulting: application to the San Francisco earthquake of 1906. Ph.D. dissertation, Univ. of Calif. at Los Angeles, Los Angeles.

Rundle, J. B. (1978). Viscoelastic crustal deformation by finite quasi-static sources. *J. Geophys. Res.* **83,** 5937–5945.

Rundle, J. B. (1980). Static elastic-gravitational deformation of a layered half-space by point couple sources. *J. Geophys. Res.* **85,** 5355–5363.

Rundle, J. B. (1982). Viscoelastic-gravitational deformation by a rectangular thrust fault in a layered earth. *J. Geophys. Res.* **87,** 7787–7796.

Rundle, J. B. (1986). An approach to modeling present-day deformation in southern California. *J. Geophys. Res.* **91,** 1947–1959.

Rundle, J. B., and Jackson, D. D. (1977a). A three-dimensional viscoelastic model of a strike slip fault. *Geophys. J. Roy. Astron. Soc.* **49,** 575–591.

Rundle, J. B. and Jackson, D. D. (1977b). A kinematic viscoelastic model of the San Francisco earthquake of 1906. *Geophys. J. Roy. Astron. Soc.* **50,** 441–458.

Ryan, J. W., Ma, C., and Caprette, D. S. (1993). NASA Space Geodesy Program—GSFC data analysis—1992. *NASA Tech. Memo*, TM-1004572.

Rybicki, K. (1971). The elastic residual field of a very long strike-slip fault in the presence of a discontinuity. *Bull. Seismol. Soc. Am.* **61,** 79–92.

Rybicki, K. (1978). Static deformation of a laterally inhomogeneous half-space by a two-dimensional strike-slip fault. *J. Phys. Earth* **26,** 351–366.

Rybicki, K., and Kasahara, K. (1977). A strike-slip fault in a laterally inhomogeneous medium. *Tectonophysics* **42,** 127–138.

Rydelek, P. A., and Sacks, I. S. (1988). Asthenospheric viscosity inferred from correlated land-sea earthquakes in north-east Japan. *Nature* **336,** 234–237.

Rydelek, P. A., and Sacks, I. S. (1990). Asthenosphere viscosity and stress diffusion: a mechanism to explain correlated earthquakes and surface deformations in NE Japan. *Geophys. J. Int.* **100,** 39–58.

Sabadini, R., and Vermeersen, L. L. A. (1997). Global post-seismic deformation: the effects of mantle and lithospheric stratification (abstract). *Eos, Trans. AGU* **78,** F155.

Sabadini, R., Yuen, D. A., and Boschi, E. (1984). The effects of post-seismic motions on the moment of inertia of a stratified viscoelastic earth with an asthenosphere. *Geophys. J. Roy. Astron. Soc.* **79**, 727–746.

Satak, K. (1993). Depth distribution of coseismic slip along the Nankai Trough, Japan, from joint inversion of geodetic and tsunami data. *J. Geophys. Res.* **98**, 4553–4565.

Sato, R. (1971). Crustal deformation due to dislocation in a multi-layered medium. *J. Phys. Earth* **19**, 31–46.

Sato, R. (1972). Stress drop for a finite fault. *J. Phys. Earth* **20**, 397–407.

Sato, R. (1974). Static deformations in an obliquely layered medium, Part I. Strike-slip fault. *J. Phys. Earth* **22**, 455–462.

Sato, R., and Matsu'ura, M. (1973). Static deformations due to the fault spreading over several layers in a multi-layered medium. Part I: Displacement. *J. Phys. Earth* **21**, 227–249.

Sato, R., and Matsu'ura, M. (1974). Strains and tilts on the surface of a semi-infinite medium. *J. Phys. Earth*, **22**, 213–221.

Sato, R., and Matsu'ura, M. (1988). A kinematic model for deformation of the lithosphere at subduction zones. *J. Geophys. Res.* **93**, 6410–6418.

Sato, R., and Matsu'ura, M. (1993). A kinematic model for evolution of island arc-trench systems. *Geophys. J. Int.* **114**, 512–530.

Sato, R., and Yamashita, T. (1975). Static deformations in an obliquely layered medium. Part II. Dip-slip fault. *J. Phys. Earth* **23**, 113–125.

Sauber, J. M., vanDam, T., Gipson, J., Herring, T., Himwich, W. E., and Clark, T. (1994). Rates of deformation southern and central Alaska from a combination of VLBI and GPS data: 1984–1993 (abstract), *Eos, Trans. AGU* **75**, 112.

Savage, J. C. (1975). Comment on 'An analysis of strain accumulation on a strike slip fault,' by D. L. Turcotte and D. A. Spence. *J. Geophys. Res.* **80**, 4111–4114.

Savage, J. C. (1980). Dislocations in seismology. In "Dislocations in Solids," Vol. 3, Moving Dislocations (F. R. N. Nabarro, ed.), North-Holland, Amsterdam, pp. 251–399.

Savage, J. C. (1983). A dislocation model of strain accumulation and release at a subduction zone. *J. Geophys. Res.* **88**, 4984–4996.

Savage, J. C. (1987). Effect of crustal layering upon dislocation modeling. *J. Geophys. Res.* **92**, 10595–10600.

Savage, J. C. (1990). Equivalent strike-slip earthquake cycles in half-space and lithosphere-asthenosphere earth models. *J. Geophys. Res.* **95**, 4873–4879.

Savage, J. C. (1995). Interseismic uplift at the Nankai subduction zone, southwest Japan, 1951–1990. *J. Geophys. Res.* **100**, 6339–6350.

Savage, J. C. (1996). Comment on 'The stress state implied by dislocation models of subduction deformation,' by J. J. Douglass and B. A. Buffett. *Geophys. Res. Lett.* **23**, 2709–2710.

Savage, J. C. (1998). Dislocation field for an edge dislocation in a layered half-space. *J. Geophys. Res.* **103**, 2439–2446.

Savage, J. C., and Burford, R. O. (1970). Accumulation of tectonic strain in California. *Bull. Seismol. Soc. Am.*, **60**, 1877–1896.

Savage, J. C., and Gu, G. (1985). A plate flexure approximation to postseismic and interseismic deformation. *J. Geophys. Res.* **90**, 8570–8580.

Savage, J. C., and Hastie, L. M. (1966). Surface deformation associated with dip-slip faulting. *J. Geophys. Res.* **71**, 4897–4904.

Savage, J. C., and Lisowski, M. (1986). Strain accumulation in the Shumagin seismic gap. *J. Geophys. Res.* **91,** 7447–7454.
Savage, J. C., and Prescott, W. H. (1978a). Asthenosphere readjustment and the earthquake cycle. *J. Geophys. Res.* **83,** 3369–3376.
Savage, J. C., and Prescott, W. H. (1978b). Comment on 'Nonlinear stress propagation in the earth's upper mantle.' *J. Geophys. Res.* **83,** 5005–5007.
Savage, J. C., and Plafker, G., (1991). Tide gauge measurements of uplift along the south coast of Alaska. *J. Geophys. Res.* **96,** 4325–4335.
Savage, J. C., and Svarc, J. L. (1997). Postseismic deformation associated with the 1992 M_w = 7.3 Landers earthquake, southern California. *J. Geophys. Res.* **102,** 7565–7577.
Savage, J. C., and Thatcher, W. (1992). Interseismic deformation at the Nankai Trough, Japan, subduction zone. *J. Geophys. Res.* **97,** 11117–11135.
Scholz, C. H., and Kato, T. (1978). The behavior of a convergent plate boundary: crustal deformation in the South Kanto District, Japan. *J. Geophys. Res.* **83,** 783–797.
Shen, Z.-K., Jackson, D. D., Feng, Y., Cline, M., Kim, M., Fang, P., and Bock, Y. (1994). Postseismic deformation following the Landers earthquake, California, 28 June 1992. *Bull Seismol. Soc. Am.* **84,** 780–791.
Shen, Z.-K., Jackson, D. D., and Ge, B. X. (1996). Crustal deformation across and beyond the Los Angeles basin from geodetic measurements. *J. Geophys. Res.* **101,** 27957–27980.
Singh, S. J. (1970). Static deformation of a multilayered half-space by internal sources. *J. Geophys. Res.* **75,** 3257–3263.
Singh, S. J., and Punia, M. (1994). A note on the effect of the free surface on the elastic deformation due to a long dip-slip fault. *J. Phys. Earth* **42,** 89–95.
Singh, S. J., and Rani, S. (1991). Static deformation due to two-dimensional seismic sources embedded in an isotropic half-space in welded contract with an orthotropic half-space. *J. Phys. Earth* **39,** 599–618.
Singh, S. J., and Rani, S. (1992). Static deformation of a uniform half-space due to a long dip-slip fault. *Geophys. J. Int.* **109,** 469–476.
Singh, S. J. and Rani, S. (1993). Crustal deformation associated with two-dimensional thrust faulting. *J. Phys. Earth* **41,** 87–101.
Singh, S. J., and Rani, S. (1994). Lithospheric deformation associated with two-dimensional strike-slip faulting. *J. Phys. Earth* **42,** 197–220.
Singh, S. J., and Rosenman, M. (1974). Quasi-static deformation of a viscoelastic half-space by a displacement dislocation. *Phys. Earth Planet. Inter.* **8,** 87–101.
Singh, S. J., Punia, M., and Rani, S. (1994). Crustal deformation due to non-uniform slip along a long fault. *Geophys. J. Int.* **118,** 411–427.
Spence, D. A., and Turcotte, D. L. (1976). An elastostatic model of stress accumulation on the San Andreas Fault. *Proc. Roy. Soc. London Ser. A* **349,** 319–341.
Spence, D. A., and Turcotte, D. L. (1979). Viscoelastic relaxation of cycle displacements on the San Andreas Fault. *Proc. Roy. Soc. London Ser. A.* **365,** 121–149.
Stein, R. S., King, G. C. P., and Rundle, J. B. (1988). The growth of geological structures by repeated earthquakes, 2, Field examples of continental dip-slip faults. *J. Geophys. Res.* **93,** 13319–13331.
Stein, R. S., King, G. C. P., and Lin, J. (1992). Changes in failure stress on the southern San Andreas fault system caused by the 1992 magnitude = 7.4 Landers earthquake. *Science* **258,** 1328–1332.
Steketee, J. A. (1958a). On Volterra's dislocations in a semi-infinite elastic medium. *Can. J. Phys.* **36,** 192–205.
Steketee, J. A. (1958b). Some geophysical applications of the elasticity theory of dislocations. *Can. J. Phys.* **36,** 1168–1197.

Stuart, W. D. (1988). Forecast model for great earthquakes at the Nankai Trough subduction zone. *Pageoph.* **126**, 619–641.

Sun, W., and Okubo, S. (1993). Surface potential and gravity changes due to internal dislocations in a spherical earth—I. Theory for a point dislocation. *Geophys. J. Int.* **114**, 569–592.

Sun, W., Okubo, S., and Vanicek, P. (1996). Global displacements caused by point dislocations in a realistic earth model. *J. Geophys. Res.* **101**, 8561–8577.

Tabei, T., Ozawa, T., Date, Y., Hirahara, K., and Nakano, T. (1996). Crustal deformation at the Nankai subduction zone, southwest Japan, derived from GPS measurements. *Geophys. Res. Lett.* **23**, 3059–3062.

Taylor, M. A. J., Zheng, G., Rice, J. R., Stuart, W. D., and Dmowska, R. (1996). Cyclic stressing and seismicity at strongly coupled subduction zones. *J. Geophys. Res.* **101**, 8363–8381.

Thatcher, W. (1974). Strain release mechanism of the 1906 San Francisco earthquake. *Science* **184**, 1283–1285.

Thatcher, W. (1975). Strain accumulation and release mechanism of the 1906 San Francisco earthquake. *J. Geophys. Res.* **80**, 4862–4872.

Thatcher, W. (1983). Nonlinear strain buildup and the earthquake cycle on the San Andreas fault. *J. Geophys. Res.* **88**, 5893–5902.

Thatcher, W. (1984). The earthquake cycle at the Nankai Trough, southwest Japan. *J. Geophys. Res.* **89**, 3087–3101.

Thatcher, W., and Rundle, J. B. (1979). A model for the earthquake cycle in underthrust zones. *J. Geophys. Res.* **84**, 5540–5556.

Thatcher, W., and Rundle, J. B. (1984). A viscoelastic coupling model for cyclic deformation due to periodically repeated earthquakes at subduction zones. *J. Geophys. Res.* **89**, 7631–7640.

Thatcher, W., Marshall, G., and Lisowski, M. (1997). Resolution of fault slip along the 470-km-long rupture of the great 1906 San Francisco earthquake and its implications. *J. Geophys. Res.* **102**, 5353–5367.

Thatcher, W., Masuda, T., Kato, T., and Rundle, J. B. (1980). Lithospheric loading by the 1896 Riku-u earthquake, Northern Japan: implications for plate flexure and asthenospheric rheology. *J. Geophys. Res.* **85**, 6429–6435.

Tse, S. T., and Rice, J. R. (1986). Crustal earthquake instability in relation to depth variation of frictional properties. *J. Geophys. Res.* **91**, 9452–9472.

Turcotte, D. L. (1992). "Fractals and Chaos in Geology and Geophysics." Cambridge Univ. Press, Cambridge.

Turcotte, D. L., and Spence, D. A. (1974). An analysis of strain accumulation on a strike-slip fault. *J. Geophys. Res.* **79**, 4407–4412.

Turcotte, D. L., Clancy, R. T., Spence, D. A., and Kulhawy, F. H. (1979). Mechanisms for the accumulation and release of stress on the San Andreas Fault. *J. Geophys. Res.* **84**, 2273–2282.

Turcotte, D. L., Liu, J. Y., and Kulhawy, F. H. (1984). The role of an intracrustal asthenosphere on the behavior of major strike-slip faults. *J. Geophys. Res.* **89**, 5801–5816.

Wahr, J., and Wyss, M. (1980). Interpretation of postseismic deformation with a viscoelastic relaxation model. *J. Geophys. Res.* **85**, 6471–6477.

Ward, S. N. (1984). A note on lithospheric bending calculations. *Geophys. J. Roy. Astron. Soc.* **78**, 241–253.

Ward, S. N. (1985). Quasi-static propagator matrices: creep on strike-slip faults. *Tectonophysics* **120**, 83–106.

Weertman, J. (1964). Continuum distribution of dislocations on faults with finite friction. *Bull. Seismol. Soc. Am.* **54,** 1035–1058.

Westerhaus, M., Woith, H., Michel, G., and Franke, P. (1997). Pore pressure induced strain anomalies (abstract). *Eos, Trans. AGU* **78,** F158.

Williams, C. A., and Richardson, R. M. (1991). A rheologically layered three-dimensional model of the San Andreas Fault in central and southern California. *J. Geophys. Res.* **96,** 16597–16623.

Yang, M., and Toksoz, M. N. (1981). Time-dependent deformation and stress relaxation after strike-slip earthquakes. *J. Geophys. Res.* **86,** 2899–2901.

Zhao, D., Christiansen, D., and Puplan, H. (1995). Tomographic imaging of the Alaska subduction zone. *J. Geophys. Res.* **100,** 6487–6504.

Zheng, G., Dmowska, R., and Rice, J. R. (1996). Modeling earthquake cycles in the Shumagins (Aleutians) subduction segment with seismic and geodetic constraints. *J. Geophys. Res.* **101,** 8383–8392.

Zienkiewicz, O. C. (1977). "The Finite Element Method," 3rd ed. McGraw-Hill, London.

INDEX

A

Accretion, 199–202
Autoregressive
  first-order
    accuracy, 10–11
    formula, 8
    power spectrum, 9–10
  multivariate, 26

B

Bandpass filters, PCA, 28–29
Boussingesq problem, 143–144

C

Carbon dioxide
  models/observations, 111–112
  northern hemisphere trends, 112–116
  spatial patterns, 111–112
CCM1, *see* NCAR community climate model
Chandler Wobble, 191
Channel, viscous flow, 181–182
CH-PCA, *see* Complex harmonic PCA
Climate, *see also* Temperatures
  oscillatory signals
    signal/noise assumptions
      actual data, comparisons, 11–13
      data set, 7
      noise component, 8–11
      signal component, 11–13
  periodic signals, 4
  seasonal cycle
    MTM-SVD analysis
      northern hemisphere trends, 112–116
      overview, 111–112
      spatial patterns, 116–120
  spatiotemporal signals
    description, 20–26
    types, 27–30
  univariate signals, 17–20
  variability
    dynamical mechanisms, 3–6
    external factors, 1–2
Complex harmonic PCA
  description, 27–28
  MTM-SVD comparison, 33
Correspondence principle
  description, 165–171
  Laplace transforms, 192
  strike-slip faults, 178
Coseismic deformation
  elastic dislocation theory
    description, 139–145
    early applications, 145–152
    finite-element method, 158–162
    lateral variations, 156–157
    spherical models, 157–158
Coulomb stresses, 195–196
Coupled earthquakes, 195–196
Creep
  deep fault, 162–163
  dislocation, 171–172
Crustal deformation
  eastern Aleutian, 210–214
  Nankai, 204–210
  overview, 204
  San Andreas fault, 214–217
  southcentral Alaska, 210–214

D

Decadal-to-century scale, 4–5
Decomposition
  PCA, 23–24
  periodic terms, 178, 180–181
  singular-value, 20–21
  steady-state terms, 178, 180–181
Deep fault creep, 162–163
Deformation
  coseismic
    elastic dislocation theory
      description, 139–145

Deformation (*Continued*)
  early applications, 145–152
  finite-element method, 158–162
  lateral variations, 156–157
  layering effects, 152–156
  matrix propagator techniques, 152–156
  spherical models, 157–158
 crustal
  eastern Aleutian, 210–214
  Nankai, 204–210
  overview, 204
  San Andreas fault, 214–217
  southcentral Alaska, 210–214
 cycle, 199
 time-dependent effects
  deep fault creep, 162–163
  description, 162
  viscoelastic flow, 163, 165
Deposition, 199–202
Dip-slip faults, 148
Dislocation creep, 171–172

E

Earthquakes, *see also specific locations*
 coupled, 195–196
 cycles
  definition, 134
  elastic half-space model, 137–139
 time-dependent effects
  accretion, 199–202
  correspondence principle, 165–171
  coupled quakes, 195–196
  deep fault creep, 162–163
  deposition, 199–202
  description, 162, 163
  erosion, 199–202
  finite-element method, 170–171
  kinematic models
   faulting environments, 183–189
   periodic terms, 178, 180–181
   steady-state terms, 178, 180–181
   subduction zone, 183–189
   three-dimensional, 182–183
   viscous flow, 181–182
  nonlinear
   flow effects, 195
   viscoelasticity, 171–172

  overview, 163
  poroelastic stress, 202–203
  postseismic rebound, 172–178
  slip, 197–199
  spherical models, 189–192
  time, 197–199
  velocity, 197–199
Elastic dislocation theory
 coseismic deformation
  description, 139–145
  early applications, 145–152
  finite-element method, 158–162
  lateral variations, 156–157
  spherical models, 157–158
Elastic half-space model, 137–139
El Niño/southern oscillation
 band signals, 66, 68–70
 CO_2 models, 121
 spatial patterns, 64–65
 SST/SLP analysis, 93–98
 typical signals, 11–12
 variability, 4
Elsasser model, 196
Empirical orthogonal function
 complex spatial patterns, 28
 extended, 27
 rotated, 27
ENSO, *see* El Niño/southern oscillation
EOF, *see* Empirical orthogonal function
Erosion, 199–202

F

Faults
 deep, creep, 162–163
 dip-slip, 148
 slip constraint, 161
 strike-slip
  correspondence principle, 166–167
  kinematic models
   faulting environments, 183–189
   periodic terms, 178, 180–181
   steady-state terms, 178, 180–181
   subduction zone, 183–189
   three-dimensional, 182–183
   viscous flow, 181–182
  long, half-space model, 137–139
  spherical models, 157
Filters, bandpass, 28–29

Finite-element method
  description, 158–162
  viscoelasticity, 170–171
Flow, nonlinear, 195
Friction, velocity-dependent, 197–199

G

Galerkin vector, 140
Geophysical fluid dynamics lab, 115–116, 120–121
GFDL, *See* Geophysical Fluid Dynamics Lab
Global Positioning System, 136
Global temperatures
  ENSO signal, 66, 68–69
  interdecadal signal, 63–65
  LFV spectra, 53–58
  overview, 51–53
  quasi-biennial signal, 69
  quasi-decadal signal, 65–66
  secular signals, 61–63
  single-grid points, 70, 72
  spatial correlations, 58–60
  temporal correlations, 58–60
Global warming, 61–62
GPS, *see* Global Positioning System
Gravity, vertical motion, 135
Green's functions
  application, 133, 135
  elastic dislocation theory, 139–144

I

Interdecadal signals
  patterns, 63–65
  SST/SLP analysis, 84–89
INTER terms, 184–185

J

Japanese quakes
  1938, 151–152
  Gifu, 151
  Nankai, 198–199, 204–210

K

Kelvin body, 165
Kinematic models
  faulting environments, 183–189
  periodic terms, 178, 180–181
  steady-state terms, 178, 180–181
  subduction zone, 183–189
  three-dimensional, 182–183
  viscous flow, 181–182

L

Landers quake, 196
Laplace transform
  description, 166
  San Andreas fault, 216
  space, 191
Layering effects, 152–156
LB94 signal
  SST patterns
    general, 84–85, 88
    NAO, 92
  time scale, 110
LFV spectra
  measuring, 36–37
  MTM-SVD analysis
    global temperatures, 53–58
    multiproxy temperatures, 104–108
    SST/SLP, 75–80
  significance, 36
Loading, gravitational, 183
Loma Prieta quake, 217
Long strike-slip fault, 137–139

M

Matrix propagator techniques, 152–156
Maxwell body
  application, 167–169
  description, 163
Models, *see specific model*
Monsoonal pattern, 88
Motion, vertical, 183–186
M-SSA, *see* Multichannel singular spectrum analysis
MTM, *see* Multiple-taper spectrum estimation methods

Multichannel singular spectrum analysis, 29–30
Multiple-taper spectrum estimation methods
  advantages, 6–7
  application, 3
  climate detection, 17–20
  description, 2–3
  –SVD
    advantages, 30–31
    application, 49–51
    conclusions, 121–122
    global temperatures
      ENSO signal, 66, 68–69
      interdecadal signal, 63–65
      LFV spectra, 53–58
      overview, 51–53
      quasi-biennial signal, 69
      quasi-decadal signal, 65–66
      secular signals, 61–63
      single-grid points, 70, 72
      spatial correlations, 58–60
      temporal correlations, 58–60
    inhomogeneity samplings, 45–49
    LFV spectrum, 38–40
    multiproxy temperatures
      LFV spectra, 104–108
      overview, 101–102
      signal reconstructions, 108–111
    seasonal cycle
      northern hemisphere trends, 112–116
      overview, 111–112
      spatial patterns, 116–120
    signal detection, 31–34
    SST/SLP analysis
      ENSO signal, 93–98
      interdecadal signal, 84–89
      LFV spectra, 75–80
      overview, 72–75
      quasi-biennial signal, 98–101
      quasi-decadal signal, 89–93
      secular signals, 80–84
    synthetic data set, 41–45
Multiproxy temperatures
  LFV spectra, 104–108
  overview, 101–102
  signal reconstructions, 108–111

N

Nankai, Japan
  earthquakes, 198–199
  subduction zone, 204–210
NAO, *see* North Atlantic oscillation
NCAR community climate model, 118, 120–121
NCAR community climate model D, 116
Noise
  background, PCA, 24
  climate
    detection assumptions, 11–13
    dynamical mechanisms, 3–6
  red, *see* Red noise
Nonlinear flow, 195
North Atlantic oscillations
  patterns
    quasi-biennial signal, 69
    quasi-decadal signal, 65–66
    signature, 68
    SLP pattern, 89–92
Northern hemisphere CO_2 trends, 112–116
Null hypothesis
  red noise, 13–14
  testing, 36–40

O

Oscillatory signals
  climate detection
    assumptions
      data set, 7
      model, 8–11
      signal factors, 11–13
    dynamical mechanisms, 3–6
    signal/noise assumptions
      actual data, comparisons, 11–13
      data set, 7
      noise component, 8–11
      signal component, 11–13
    variability, 1–2
  ENSO
    band, 66, 68–70
    CO_2 models, 121
    MTM-SVD analysis, 68–69
    spatial patterns, 64–65
    SST/SLP analysis, 93–98
    typical, 11–12
    variability, 4
  interdecadal
    description, 63–65
    SST/SLP analysis, 84–89

INDEX

multiproxy temperatures, 108–111
principal patterns, 26
quasi-decadal
  global temperatures, 65–66
  SST/SLP analysis, 89–93
secular, 61–63, 80–84
single-grid points, 70, 72
time-domain, 63–64

P

PCA, see Principal component analysis
Periodic terms, 178, 180–181
POPs, see Principal oscillation patterns
Pore pressure, 202–203
Poroelasticity, 202–203
Postseismic rebound, 172–178
POST terms, 184–185
Power spectrums, AR(1), 9–10
Pressure, pore, 202–203
Prince William Sound quake
  deep fault creep, 163
  dislocation model, 210–211
  dislocation theory, 148–149
  viscoelastic model, 213–214
  VLBI, 192
Principal component analysis
  application, 22–26
  bandpass-filter, 28–29
  classical time-domain, 62
  complex harmonic, 27–28
  decomposition, 23–24
  MTM-SVD analysis, 43–44
  spectral analysis, 20–22
Principal oscillation patterns, 26

Q

Quasi-decadal signals, SST/SLP analysis, 89–93

R

Rebound, postseismic, 172–178
Red noise
  AR(1), 8–11
  modes, 3
  null hypothesis, 13–14

S

San Andreas fault
  crustal deformation, 214–217
  deformation, 182
  three-dimensional model, 182–183
San Fernando quake, 151
San Francisco quakes
  1906
    dislocation theory, 151–152
    postseismic rebound, 173, 175
  Loma Prieta, 217
San Jacinto fault, 195
Sea-level pressure
  ENSO signal, 93–98
  interdecadal signal, 84–89
  LFV spectra, 75–80
  overview, 72–75
  quasi-biennial signal, 98–101
  quasi-decadal signal, 89–93
  secular signals, 80–84
Seasonal cycle
  northern hemisphere trends, 112–116
  overview, 111–112
  spatial patterns, 116–120
Secular signals
  SST/SLP analysis, 84–89
  temperature analysis, 61–63
Shumagin Islands, 188–189
Signals
  amplitudes, 12–13
  El Niño/southern oscillation, 11–12
  MTM-SVD detection, 31–34
  oscillatory, see Oscillatory signals
  periodic, 4
  spatiotemporal, 41–45
Singular-value decomposition
  –MTM
    advantages, 30–31
    application, 49–51
    conclusions, 121–122
    global temperatures
      ENSO signal, 66, 68–69
      interdecadal signal, 63–65
      LFV spectra, 53–58
      overview, 51–53
      quasi-biennial signal, 69
      quasi-decadal signal, 65–66
      secular signals, 61–63
      single-grid points, 70, 72

Singular-value decomposition (*Continued*)
    spatial correlations, 58–60
    temporal correlations, 58–60
    inhomogeneity samplings, 45–49
    LFV spectrum, 38–40
    multiproxy temperatures
        LFV spectra, 104–108
        overview, 101–102
        signal reconstructions, 108–111
    seasonal cycle
        northern hemisphere trends, 112–116
        overview, 111–112
        spatial patterns, 116–120
    signal detection, 31–34
    SST/SLP analysis
        ENSO signal, 93–98
        interdecadal signal, 84–89
        overview, 72–75
        quasi-biennial signal, 98–101
        quasi-decadal signal, 89–93
        secular signals, 80–84
    synthetic data set, 41–45
    –PCA, application, 20–21
Slip occurrence, 197
SLP, *see* Sea-level pressure
Spatiotemporal signals, 41–45
Spectral analysis
    PCA and, 20–26
    rotated EOFs and, 27
Spherical models
    elastic properties, 157–158
    time-dependent, 189–192
Steady-state terms, 178, 180–181
Stress
    change, 196
    Coulomb, 195–196
    poroelastic, 202–203
Strike-slip faults
    kinematic models
        description, 178, 180–181
        faulting environments, 183–189
        periodic terms, 178, 180–181
        steady-state terms, 178, 180–181
        subduction zone, 183–189
        three-dimensional, 182–183
        viscous flow, 181–182
    spherical model, 157
Subduction zones
    eastern Aleutian, 210–214
    kinematic models, 183–189
    Nankai, 204–210
    southcentral Alaska, 210–214
Surface temperature
    MTM-SVD analysis
        ENSO signal, 93–98
        interdecadal signal, 84–89
        overview, 72–75
        quasi-biennial signal, 98–101
        quasi-decadal signal, 89–93
        secular signals, 80–84
SVD, *see* Singular-value decomposition

T

Tanna fault, 148
TECTON, 175–177
Temperatures, *see also* Climate
    global
        MTM-SVD analysis
            ENSO signal, 66, 68–69
            interdecadal signal, 63–65
            LFV spectra, 53–58
            overview, 51–53
            quasi-biennial signal, 69
            quasi-decadal signal, 65–66
            secular signals, 61–63
            single-grid points, 70, 72
            spatial correlations, 58–60
            temporal correlations, 58–60
    multiproxy
        LFV spectra, 104–108
        overview, 101–102
        signal reconstructions, 108–111
    surface
        MTM-SVD analysis
            ENSO signal, 93–98
            interdecadal signal, 84–89
            overview, 72–75
            quasi-biennial signal, 98–101
            quasi-decadal signal, 89–93
            secular signals, 80–84
Tide gauges, 211
Time-dependent effects
    accretion, 199–202
    coupled earthquakes, 195–196
    deep fault creep, 162–163
    deposition, 199–202
    description, 162, 163
    erosion, 199–202

finite-element method, 170–171
kinematic models
    faulting environments, 183–189
    periodic terms, 178, 180–181
    steady-state terms, 178, 180–181
    subduction zone, 183–189
    three-dimensional models, 182–183
    viscous flow, 181–182
nonlinear, 171–172
nonlinear flow, 195
poroelastic stress, 202–203
postseismic rebound, 172–178
slip, 197–199
spherical models, 189–192
time, 197–199
velocity, 197–199
viscoelastic flow, 165–171
Time-domain signals, 63–64

U

Univariate signal detection, 17–20
Uplift occurence, 175–176

V

Variability
    climate
        dynamical mechanisms, 3–6
        external factors, 1–2
    decadal-to-century scale, 4–5
    El Niño/southern oscillation, 4
Velocity-dependent friction, 197–199
Vertical motion, 183–186
Very long baseline interferometry, 192
Virtual work principle, 158–160
Viscoelasticity
    model, 213–214
    poroelasticity vs., 203
Viscous flow, 181–182
VLBI, see Very long baseline interferometry

W

Western Pacific oscillation, 68
WPO, see Western Pacific oscillation

ISBN 0-12-018841-4